LONG-TERM MANAGEMENT OF CONTAMINATED SITES

RESEARCH IN SOCIAL PROBLEMS AND PUBLIC POLICY

Series Editors: William R. Freudenburg and
Ted I. K. Youn

Recent Volumes:

Volume 8: The Organizational Response to Social Theory – Edited by Stephanie W. Hartwell and Russell K. Schutt, 2000

Volume 9: Environmental Risks: Perception, Evaluation and Management – Edited by G. Böhm, J. Nerb, T. McDaniels and H. Spada, 2001

Volume 10: The Environmental State under Pressure – Edited by Arthur P.J. Mol and Frederick H. Buttel, 2002

Volume 11: Terrorism and Disaster: New Threats, New Ideas – Edited by Lee Clarke, 2003

Volume 12: The Organizational Response to Persons with Mental Illness Involved with the Criminal Justice System – Edited by Stephanie W. Hartwell, 2005

RESEARCH IN SOCIAL PROBLEMS AND PUBLIC POLICY
VOLUME 13

LONG-TERM MANAGEMENT OF CONTAMINATED SITES

EDITED BY

THOMAS M. LESCHINE
School of Marine Affairs, University of Washington, USA

Emerald

United Kingdom – North America – Japan
India – Malaysia – China

Emerald Group Publishing Limited
Howard House, Wagon Lane, Bingley BD16 1WA, UK

Reprints and permission service
Contact: booksandseries@emeraldinsight.com

British Library Cataloguing in Publication Data
A catalogue record for this book is available from the British Library

ISBN: 978-0-7623-1319-8
ISSN: 0196-1152 (Series)

Printed and bound by CPI Group (UK) Ltd, Croydon, CR0 4YY

ISOQAR QUALITY ASSURED ISO 9001 UKAS QUALITY MANAGEMENT 026
Certificate number 1985

Awarded in recognition of Emerald's production department's adherence to quality systems and processes when preparing scholarly journals for print

INVESTOR IN PEOPLE

CONTENTS

LIST OF CONTRIBUTORS *ix*

INTRODUCTION: LONG-TERM MANAGEMENT OF CONTAMINATED SITES
Thomas M. Leschine *1*

PUBLIC PARTICIPATION IN LONG-TERM MANAGEMENT

THE HANFORD OPENNESS WORKSHOPS: FOSTERING OPEN AND TRANSPARENT LONG-TERM DECISION MAKING AT THE DEPARTMENT OF ENERGY
Christina H. Drew, Michael Kern, Todd Martin, Max S. Power and Elaine M. Faustman *13*

COMPETING PERSPECTIVES ON A PROCESS FOR MAKING REMEDIATION AND STEWARDSHIP DECISIONS AT THE ROCKY FLATS ENVIRONMENTAL TECHNOLOGY SITE
Seth Tuler and Thomas Webler *49*

LONG-TERM MANAGEMENT IN PRACTICE: LESSONS FROM THE SITES

LONG-TERM STEWARDSHIP FOR THE HANFORD NUCLEAR SITE
Max S. Power *81*

CERCLA REMEDY SELECTION LESSONS: THE SHATTUCK SITE
Elizabeth K. Hocking 105

SCIENTIFIC AND TECHNICAL PERSPECTIVES ON LONG-TERM MANAGEMENT

A SUSTAINABLE ENVIRONMENTAL PROTECTION SYSTEM FOR THE MANAGEMENT OF RESIDUAL CONTAMINANTS
Kevin M. Kostelnik, James H. Clarke, Jerry L. Harbour, Florence Sanchez and Frank L. Parker 117

ECOLOGICAL CONSIDERATIONS FOR EVALUATING CURRENT RISK AND DESIGNING LONG-TERM STEWARDSHIP ON DEPARTMENT OF ENERGY LANDS
Joanna Burger, Nellie Tsipoura, Michael Gochfeld and Michael R. Greenberg 139

THE LESSONS OF HISTORY

LIVING IN A NUCLEAR LANDSCAPE: REHABILITATION AND RESETTLEMENT OF PROVING GROUNDS IN AUSTRALIA AND ISLANDS OF THE WESTERN PACIFIC
Anne Ballou Jennings, Amy M. Seward and Thomas M. Leschine 165

THE CHALLENGES AHEAD

VULNERABILITIES AND UNCERTAINTIES IN LONG-TERM STEWARDSHIP
Curtis Travis 195

A 19-YEAR PERSPECTIVE ON LONG-TERM CARE ISSUES
Denise Bierley *213*

LONG-TERM STEWARDSHIP AND RISK MANAGEMENT: ANALYTIC AND POLICY CHALLENGES
Eugene A. Rosa *227*

LIST OF CONTRIBUTORS

Denise Bierley	Outrider Environmental Planning and Technical Services, Inc., Albuquerque, NM, USA
Joanna Burger	Division of Life Sciences Nelson Labs, Rutgers University Piscataway, NJ, USA
James H. Clarke	Department of Civil and Environmental Engineering, Vanderbilt University, Nashville, TN, USA
Christina H. Drew	Ross & Associates Environmental Consulting, Ltd., Seattle, WA, USA
Elaine M. Faustman	Institute for Risk Analysis and Risk Communication, Department of Environmental and Occupational Health Sciences, University of Washington, Seattle, WA, USA
Michael Gochfeld	Environmental and Community Medicine, UMDNJ-Robert Wood Johnson Medical School, Piscataway, NJ, USA
Michael R. Greenberg	E. J. Bloustein School of Planning and Policy, Rutgers University, New Brunswick, NJ, USA
Jerry L. Harbour	Idaho National Laboratory, Idaho Falls, ID, USA
Elizabeth K. Hocking	Argonne National Laboratory, Washington, DC, USA
Anne Ballou Jennings	Consultant, Seattle, WA, USA
Michael Kern	Long Live the Kings, Seattle, WA, USA

Kevin M. Kostelnik	Idaho National Laboratory, Idaho Falls, ID, USA
Thomas M. Leschine	School of Marine Affairs, University of Washington Seattle, WA, USA
Todd Martin	Consultant, Northport, WA, USA
Frank L. Parker	Department of Civil and Environmental Engineering, Vanderbilt University, Nashville, TN, USA
Max S. Power	Washington Department of Ecology, (retired), Lacey, WA, USA and Member, Oregon Hanford Cleanup Board, Yachats, OR, USA
Eugene A. Rosa	Department of Sociology, Thomas F. Foley Institute for Public Policy & Public Service, Washington State University, Pullman, WA, USA
Florence Sanchez	Department of Civil and Environmental Engineering, Vanderbilt University, Nashville, TN, USA
Amy M. Seward	Global Security Technology & Policy Group, Pacific NW National Lab, Seattle, USA
Curtis Travis	Science Applications International Corporation (SAIC), Knoxville, TN, USA
Nellie Tsipoura	New Jersey Audubon Society, Bernardsville, NJ, USA
Seth Tuler	Social and Environmental Research Institute, Greenfield, MA, USA
Thomas Webler	Social and Environmental Research Institute, Greenfield, MA, USA and Department of Environmental Studies, Antioch University New England, Keene, NH, USA

INTRODUCTION: LONG-TERM MANAGEMENT OF CONTAMINATED SITES

Thomas M. Leschine

A recent National Research Council study estimates that there are now 217,000 contaminated sites in the United States (NRC, 2003a). The proliferation of hazardous contamination across the landscape is an unwelcome if unsurprising byproduct of industrialization during the past century and the ledger continues to grow despite billions spent on remediation. Both government and the private sector are culpable in the production and disposal practices that created these sites. Although most sites are small and privately owned, the largest, and the majority of the most hazardous sites, were created by government itself. This is particularly the case with respect to nuclear weapons production, development and testing, but is also the result of other defense-related activities. These sites collectively contain billions of cubic yards of soil and groundwater in need of remediation (NRC, 2003a). Many would threaten both the environment and human health in their current condition, if present-day management control were to be neglected or lost.

A pair of previous National Research Council reports (NRC, 2000, 2003b) addressed radioactively contaminated sites managed by the U.S. Department of Energy (DOE). The NRC studies found that most of these "legacy" sites (as they are now referred to) will not be cleaned up to the point where they can support unrestricted access; that most will require care

Long-Term Management of Contaminated Sites
Research in Social Problems and Public Policy, Volume 13, 1–10

ISSN: 0196-1152/doi:10.1016/S0196-1152(06)13012-4

into the indefinite future because of the residual risks; that the needs of what has come to be called "long-term stewardship" are not being adequately taken into account in current site planning; and that the lesson of history is that neither engineered controls nor institutional management measures can be counted upon to remain effective for as long as many of the most dangerous contaminants will remain.

The two NRC studies that produced these findings are the impetus for this volume. Several of the authors who appear in these pages, including series editor William Freudenburg and volume editor Thomas Leschine, were members of one or the other of the two committees that authored these reports. Service by authors on these and other NRC panels dealing with nuclear or other long-lived hazards is thus a common if invisible thread that ties many of these chapters together, and reports of the National Research Council are frequently cited by this volume's contributors. The perspective is interdisciplinary, as is the perspective of the two NRC reports. The authors come from the natural and social sciences and engineering, occupational and environmental health, or backgrounds in law, regulatory decision making or public policy. Some are academics, some practitioners, and some have been both over long careers. Some speak from experience "in the trenches" as regulators, as community leaders in the public process of weighing cleanup alternatives in light of hoped-for site futures, or as managers, consultants or technical specialists who have worked for governmental waste management programs.

The prospects for effective institutional management for contaminated sites – both today and over the very long time periods that are necessary in some cases – are the principal concern of this volume. Readers will discover that neither "remediation" nor "cleanup" – terms in common use in present-day waste site management – necessarily follow the customary definition of "cleaned up". As Curtis Travis points out (*Vulnerabilities and Uncertainties in Long-Term Stewardship*), sites are frequently declared "cleaned up and closed" not because contaminants have been removed or permanently isolated from the biotic environment, but instead because engineered controls to prevent waste migration, and institutional controls to regulate human access and use, have been established in accordance with federal and state law and regulation, other organizational imperatives and/or procedural guidance of the managing entities.

Kevin Kostelnik and colleagues (*A Sustainable Environmental Protection System for the Management of Residual Contaminants*) argue that the long-term protectiveness of engineered containment barriers for hazardous wastes left in environmental media is determined not so much by physical parameters of design and construction as by the extent to which the planning for design, construction and on-going surveillance and maintenance is an integral part of fully realized remediation and long-term stewardship planning. Current site planning is seldom adaptive in the ways these authors (and all three NRC studies cited above) argue that it needs to be, however, typically dominated instead by linear thinking and regulatory-driven decisions. Both Curtis Travis, mentioned above, and Denise Bierley (*A Nineteen Year Perspective on Long-Term Care Issues*) point to this disjunction in their contributions to this volume's final summative section (*The Challenges Ahead*). Gene Rosa goes a step further in the volume's concluding chapter (*Long-term Stewardship and Risk Management: Analytic and Policy Challenges*), critiquing from first principles the organizational and intellectual bases for the heuristics that organizations engaged in management of high-risk technologies typically employ. Denise Bierley, writing from the perspective of many years in various practitioner roles, argues that the hoped-for evolution in thinking can occur if managers, policy makers and local communities develop a shared vision of problems and solutions and, as an example, points to aspects of DOE's Uranium Mill Tailings Remedial Action (UMTRA) Program, one that addresses 24 sites scattered across 10 states.

The sites that are the focus of this volume will not be cleaned up to the extent we had expected a generation ago when landmark laws like "Superfund" (the Comprehensive Environmental Response, Compensation and Liability Act (CERCLA) of 1980) were passed. Rapidly rising cleanup costs are prominent among the reasons, in part because of their direct relationship with the cleanup standards applied, in turn a function of the cleanup's target "end state". Curtis Travis makes the point that federal agencies increasingly claim to focus on starting with the target end state to determine the required cleanup level, seemingly a step with compelling rationality, but in reality the cleanup level seems to be dictated by costs and other considerations. Host communities aspire not to be burdened with site cleanups that limit future options once the DOE mission is ended, a continuing source of intergovernmental conflict.

Local communities have numerous avenues by which they can attempt to influence the character of cleanup and long-term management for nearby sites. The variety of roles and outcomes achieved through community-regulator interaction is highlighted in the first four chapters of this volume – Christina Drew and colleagues on the Hanford Site's "openness" workshops that grew out of stakeholder initiative (*The Hanford Openness Workshops: Fostering Open and Transparent Long-Term Decision-Making at the Department of Energy*); Seth Tuler and Thomas Webler on preferences across stakeholder groups with respect to the processes employed for deciding cleanup at the Rocky Flats Site (*Competing Perspectives on a Process for making Remediation and Stewardship Decisions at the Rocky Flats Environmental Technology Site*); Max Power on the state's role in influencing DOE's charting of a cleanup strategy for Hanford (*Long-Term Stewardship for the Hanford Nuclear Site*); and Elizabeth Hocking on community aspirations and attitudes with respect to remedy selection at the Shattuck Site, located within the city limits of Denver, Co. (*CERCLA Remedy Selection Lessons: The Shattuck Site*).

What we mean by "sites" in the description of the basic problem that motivates this volume are those sites and situations that, regardless of how they were created, are now under a semblance of administrative control: guardianship if not cleanup-in-progress. Kevin Kostelnik and co-authors describe modern management practices for sites like these. Some of the largest and most contaminated sites are actually comprised of "sites within sites", particularly the vast nuclear reservations – Hanford (Washington), Savannah River (Georgia), Rocky Flats (Colorado), and the Idaho National Laboratory, among others. These reservations, and other sites referred to collectively as "the weapons complex", were created during World War II or thereafter for the purpose of nuclear weapons development, production, assembly and testing.

The Hanford Nuclear Reservation, the archetype of this class of sites, is by most measures the most contaminated and now the locus of the most expensive environmental cleanup operation on the planet (costing between $6 and $8 billion per year in recent years) . Originally, some 560 square miles in extent, Hanford's effective "footprint" has shrunk as cleanup has progressed. Much land has already become wildlife refuges, land that remained relatively uncontaminated throughout the period of site operations, when it served mainly to provide defense perimeters for operations conducted under high levels of government-imposed secrecy.

Such buffer areas are today lands "frozen in time", and their existence presents one of several socio-cultural dilemmas that are encountered in this volume. What should their use be once a site's original mission has ended? Most are now extremely valuable: for the *de facto* ecological protection they provide, for their development potential as the last large undeveloped spaces in geographical areas otherwise surrounded by creeping suburbanization, for their importance to the cultural identity of the tribal, aboriginal, or other original people who willingly vacated ancestral lands in a time of national need, and for their continued value as buffers given that the contamination at the core requires continued restrictions on human presence. Joanna

Burger and her colleagues ask whether we should clean up at all if to do so is to invite human conversion of what would otherwise be valuable ecological areas. They note that important ecological attributes of some sites will be put at risk by the very act of cleaning up, as might be worker health and safety (*Ecological Considerations for Evaluating Current Risk and Designing Long-Term Stewardship on Department of Energy Lands*).

The weapons testing ranges considered in this volume exemplify another set of challenges to institutional management, as their present-day legacies often include extensive areas of surface contamination in lands that, until the testing began, had been occupied by indigenous people (Anne Jennings and colleagues, *Living in a Nuclear Landscape: Rehabilitation and Resettlement of Proving Grounds in Australia and Islands of the Western Pacific*). Present day resettlement efforts at both Australian and Pacific island sites underscore the difficult social and cultural questions that arise when a land's original people desire to resume a lifestyle that was abruptly interrupted in a time of national need but whose resumption today, in light of lingering site contamination, pits human health risk against culture. Some large U.S. DOE western sites raise similar questions, since they occupy lands either ceded by Indian Tribes or that otherwise continue to hold cultural and economic significance as spiritual or hunting and gathering sites to the tribes that have continued to occupy lands nearby.

The broader lesson is that site condition and the social condition of adjacent communities are often inextricably linked. Although receiving comparatively little formal attention in law and policy, linkages between culture and contamination often underlie political conflict over site cleanup. These influences are characteristic of the final category of sites considered in this volume – those that require us to continue to be able to live with them as we have in the past, because they are and always have been "part of the neighborhood", so to speak, perhaps just down the road from where we live. Many of the Superfund sites that now dot the American landscape fall into this category. The Shattuck Site occupies just 10 acres within the city limits of Denver, Colorado, but a cost-minimizing approach to cleanup left significant quantities of waste in place and requires extensive reliance on institutional controls for long-term protection. The bargain initially struck was soon being questioned by a skeptical community (Elizabeth Hocking, *CERCLA Remedy Selection Lessons: The Shattuck Site*).

The long-term nature of the risks described above – and the various ways in which technological, ecological, socio-cultural, and institutional systems are interconnected in their production and management – are the central concerns of this volume. In the absence of cleanup to the point where unrestricted use becomes possible, a status that is increasingly unlikely for most sites, the hazards that exist at many sites will outlive both engineered and organizational systems put into place to deal with them. To be sure, there are benefits to continued use-restrictions in some cases – more assured ecological protection as Burger and colleagues suggest, a more rapid return of land or facilities to productive industrial use that does not require the same degree of protection as unrestricted use, or perhaps benefits to society at large that come from avoided costs of additional cleanup.

The lesson of the past, however, is not only that effective long-term management cannot be assured but also that intended long-term measures may fail in the short run. A touchstone in several of the chapters that follow is the infamous Love Canal incident. It serves as a bellwether for public and governmental concern in this arena, having been a major catalyst for the Superfund law passed in 1980. The Love Canal episode began to unfold in the 1950s with the construction of an elementary school directly atop what had until a few years earlier been a dumping ground for thousands of tons of extremely hazardous chemicals. Soon a neighborhood followed, under the dual attraction of the school and inexpensive land. Land transfers, initially from the private to the public sector, unwittingly aided the growth of a local population at risk from hazards that were not recognized and by the late 1970s a pattern of illnesses had developed in the adjacent neighborhood that could no longer be ignored. Curtis Travis brings the Love Canal story up to date, pointing out that although a much more protective technical solution has been put into place, 21,000 tons of chemical wastes remain on site, and efforts are underway to renovate and expand the formerly contaminated housing that was abandoned some 25 years ago. The final EPA decision documents were issued in 2003. The path from problem discovery to problem resolution spanned nearly 25 years, but the decision by the local board of education to build an elementary school on land that had been transferred with, by many accounts, warnings about the hazards that lay below occurred within about a year of the land transfer (Freudenburg in NRC, 2000, p. 53).

Kostelnik and colleagues provide us with a more recently uncovered example of such "institutional forgetting", closing their chapter with the tale of the discovery, in 1993, of a long-forgotten World War I munitions manufacturing site that is the present-day location of both American University and an adjacent upscale neighborhood in Washington, D.C. Long-forgotten munitions were uncovered during construction of a private residence on land that had been deeded originally by the Army to American University, but subsequently sold for private development. The cost of site remediation is estimated to be in excess of $100 million. Land transfers emerge repeatedly as the "Achilles' heel" of long-term care intentions, due in part to inherent conflict between the desires of local jurisdictions for redevelopment and the mandates of higher authorities that the institutional memory of buried contamination not be extinguished (NRC, 2000, Chapter 7 and Appendix E). It is fair to ask, as physicist and science fiction writer Gregory Benford (1999) does in his provocative book *Deep Time* (1999), how people in the distant future will even recognize that we are trying to tell them we have left behind something that in our time was regarded as hazardous.

Iterative, adaptive and learning-oriented strategies that probe constantly to detect and correct weakness and error are the forms of long-term stewardship most likely to succeed (NRC, 2000, 2003b), but how can we assure that the institutions that manage hazards will retain these characteristics through time, even if they could produce them in the short run? The problem takes on attributes of infinite regression, and the desired meta-characteristics of the institutional designs we seek seem to come down to high levels of public trust and reliability, and the ability to maintain constancy of purpose through time (Keller & La Porte, 1996; NRC, 2003b; La Porte, 2004, amplified in chapters by Bierley and Rosa). Past studies of the institutional proclivities that impede the hoped-for designs have generated a rich vocabulary for diagnosing institutional and organizational maladies – "the disqualification heuristic" (Clarke, 1993), "the atrophy of vigilance" (Freudenburg, 1992), "the rhetoric of certitude" (Rosa & Clark, 1999) and "institutional recreancy" (Freudenburg, 1993), among others.

The papers in this volume are presented in five sections. The volume opens with a focus on public participation in waste management decision making that has echoes in the chapters that follow. Drew and colleagues detail the history of operations and disposal practices at the Hanford Site in a way that makes this chapter a useful introduction for those unfamiliar with the defense wastes problem arena. The success of the stakeholder-driven Hanford Openness Workshops in creating a window into site operations and records that had been shrouded in secrecy for decades provides a lesson that is at once both positive and cautionary, as many of the gains made were quickly lost in the aftermath of the 9/11 attacks. Tuler and Webler examine the controversy over the choice of a technical standard to guide the cleanup of surface soils contaminated with plutonium residues at the Rocky Flats site not as a question of preferences for outcomes but rather as a question of whether there is a "best" decision making and public participation process to follow in engaging the public in such decisions. Their use of Q method leads to a characterization of distinct perspectives on decision-making approaches, with their conclusion a challenge to the presumption that there is "a" decision-making approach that is a priori right for all circumstances.

The next section provides two reflections on the strategies and outcomes of decision processes at two radioactively contaminated sites at polar opposites in the site classification presented above – the 560 sq mi Hanford Site located in the semi-arid reaches of the Columbia River plateau in eastern Washington State, formerly the nation's principal plutonium production factory, and the 10 acre Shattuck site located in an industrial area of Denver, Colorado. The latter typifies the processing, fabrication and other small manufacturing facilities within the weapons complex that were scattered across cities and towns throughout the United States during the era of the Cold War. Max Power's and Elizabeth Hocking's examinations of attempts to define cleanup and long-term stewardship protocols for these very different sites nevertheless reveal common threads, notably similar institutional shortcomings and regulatory mismatch with similar ramifications for agency credibility and public trust.

The third section provides the perspectives of two groups of technical experts, Kevin Kostelnik and colleagues, a team of civil and environmental engineers and chemists, and Joanna Burger and colleagues, ecologists, an occupational health specialist, and a policy specialist. Each group speaks to the long-term stewardship construct from their disciplinary perspectives as well as extensive field work at various DOE sites.

The fourth section, consisting of a single paper that takes a historical perspective, provides an illuminating comparison of the attempts of two indigenous cultures – the Micronesian people of the Marshall Islands and the Maralinga people (the Maralinga Tjarutja), the indigenous people of South Australia – to reclaim lands they ceded to the American and British governments, respectively, for nuclear weapons testing. The two cultures today exist in very different political relationships with the larger governmental authorities they are trying to influence, and they are pursuing different strategies as a result. But both seem to have arrived at a place in their efforts where they face common questions of how and whether long-held cultural practices can co-exist with long-lasting environmental contamination of very modern origin.

The final section brings us to broad reflections on the prospects for long-term management from the perspectives of a risk and systems analyst with experience in numerous environmental risk arenas (Travis), a practitioner who has worked both as a technical consultant and a state regulator (Bierley), and a sociologist whose specialties include organizational aspects of the management of high-risk technologies (Rosa).

The terms "long-term management", "long-term care" and "long-term stewardship" first emerged in the context of long-lived hazards like those discussed in this volume during the 1990s, with the first governmental efforts to estimate the financial liability for the environmental legacy of the Cold War. The idea that these wastes would not be cleaned up, financial and technical limitations notwithstanding, represents a substantial political and cultural shift that is still underway. The papers in this volume, and the other work upon which these ideas are built, suggest that to move "from cleanup to stewardship", as the title of one prominent DOE report proposes (U.S. DOE, 1999), is to move into very different territory indeed, both operationally and conceptually. As Todd La Porte (2004) has remarked, the organizational and management challenges are extraordinary, in many ways unprecedented. The key government agencies in whose care these wastes are have been slow to acknowledge these realities. With the publication of this volume there is some basis for asserting that we now at least know what the questions are. The search for workable institutional designs that lead to robust systems of long-term care will have to continue.

ACKNOWLEDGMENTS

This volume would not have been possible without the efforts of the anonymous peer-reviewers of the individual chapters and the invaluable editorial and organizational assistance of Kit O'Neill, to all of whom the editor offers his thanks. Sue Latourell and Amy Seward of the School of Marine Affairs at the University of Washington assisted during the early stages of volume preparation.

The volume had its origins in committee work of the National Research Council on long-term stewardship of nuclear wastes, in which several of the authors participated. Ideas and collaborations were further developed in professional meetings of the Society for Risk Analysis in Baltimore in December, 2003, and the American Association for the Advancement of Science in Seattle in February, 2004.

REFERENCES

Benford, G. (1999). *Deep time: How humanity communicates across millennia.* New York: William Morrow.

Clarke, L. B. (1993). The disqualification heuristic: When do organizations misperceive risk? *Research in Social Problems and Public Policy, 5*, 289–312.

Freudenburg, W. R. (1992). Nothing recedes like success? Risk analysis and the organizational amplification of risks. *Risk, 3*(1-Winter), 1–35.

Freudenburg, W. R. (1993). Risk and recreancy: Weber, the division of labor and the rationality of risk perceptions. *Social Forces, 71*(4), 909–932.

Keller, A., & La Porte, T. R. (1996). Assuring institutional constancy: Requisite for managing long-lived hazards. *Public Administration Review, 56*(6), 535–544.

La Porte, T.R. (2004). Elements for long-term institutional stewardship in a hazardous age. Paper presented at Annual Meeting of the American Association for the Advancement of Science, Seattle, Washington.

National Research Council (NRC). (2000). *Long-term institutional management of U.S. department of energy legacy waste sites*. Washington, DC: National Academies Press.

NRC. (2003a). *Environmental cleanup at navy facilities: Adaptive site management*. Washington, DC: National Academies Press.

NRC. (2003b). *Long term stewardship of DOE legacy waste sites – A status report*. Washington, D.C.: National Academies Press.

Rosa, E. A., & Clark, D. L., Jr. (1999). Historical routes to technological gridlock: Nuclear technology as prototypical vehicle. *Research in Social Problems and Public Policy, 7*, 21–57.

U.S. Department of Energy (DOE). (1999). *From cleanup to stewardship*. A Companion Report to Accelerating Cleanup: Paths to Closure. Office of Environmental Management DOE/EM-0466, Washington, DC.

PUBLIC PARTICIPATION IN LONG-TERM MANAGEMENT

THE HANFORD OPENNESS WORKSHOPS: FOSTERING OPEN AND TRANSPARENT LONG-TERM DECISION MAKING AT THE DEPARTMENT OF ENERGY

Christina H. Drew, Michael Kern, Todd Martin, Max S. Power and Elaine M. Faustman

ABSTRACT

Openness is critical to long-term cleanup and stewardship of former nuclear weapons manufacturing facilities. Openness is especially challenging in such facilities because a culture of secrecy has dominated them since the earliest days of nuclear weapons development. This paper describes a multi-year effort at the Hanford Site called the Hanford Openness Workshops. The Workshops were convened to address and resolve issues impeding the availability and understanding of information important to public health, the environment, and decision–making. Lessons from the workshops can improve dialogue about nuclear waste cleanup and long-term stewardship among regulators, decision makers, stakeholders, and Tribes.

Long-Term Management of Contaminated Sites
Research in Social Problems and Public Policy, Volume 13, 13–48

ISSN: 0196-1152/doi:10.1016/S0196-1152(06)13001-X

INTRODUCTION

For nearly 50 years, plutonium for nuclear weapons was produced at the Hanford site, located in southeastern Washington State, near the city of Richland (Fig. 1). Construction at Hanford began in 1943 as part of the Manhattan Project. Hanford's "B Reactor" produced the plutonium used in the bomb dropped on Nagasaki, Japan at the end of World War II. Today, Hanford is owned by the United States Department of Energy (DOE) and managed by the DOE-Richland Operations Office (DOE-Richland). Producing materials for nuclear weapons has left an enormous legacy of radioactive and chemical wastes at Hanford. DOE-Richland currently employs several major private contractors who are conducting a decades-long cleanup effort that will cost 50–60 billion dollars (Dininny, 2003). Even after cleanup is "complete," long-term oversight of Hanford and other former nuclear weapons facilities will be necessary to monitor and address residual contamination.

Nuclear weapons development occurred within a strong culture of secrecy. The secrecy was necessary, first to prevent the Axis powers from knowing the United States had developed an atomic bomb, and later to prevent the

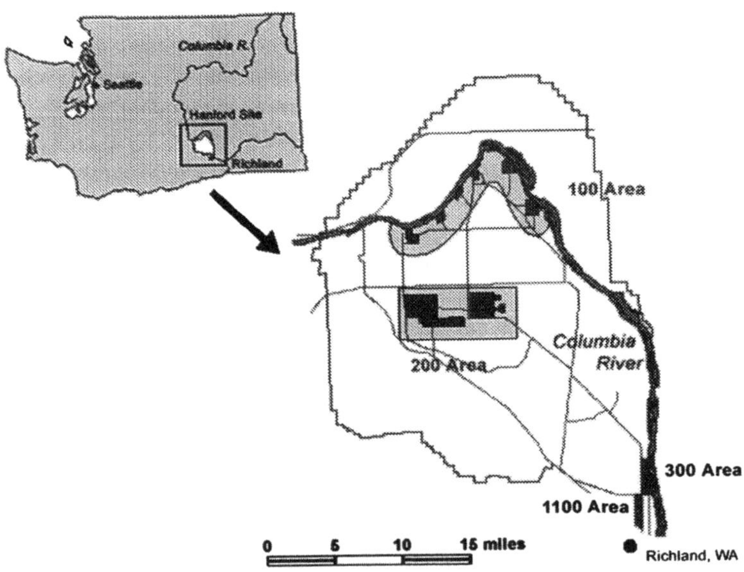

Fig. 1. The Hanford Site in Washington State, USA.

spread of nuclear weapons technology to other countries. During the Cold War, virtually all documents associated in any way with the production and development of nuclear weapons were classified. Government officials believed the information contained in these documents – if released – could jeopardize the security of the nation. National security was also used as a rationale for hiding environmental, health, and safety documentation from the public. Officials were concerned about public reaction if the environmental impacts of weapons production were disclosed (Gordon et al., 1973; Parker, 1948). After the Cold War, the mission at Hanford changed from weapons production to waste management and cleanup. This, coupled with rising public awareness and influence about environmental contamination at the site, created pressure for a more open and participatory decision process at Hanford and other weapons sites. Today, the increasing national focus on the long-term stewardship of the weapons complex requires special attention to the issue of "openness," since it remains a significant challenge for all DOE sites.

The purpose of this paper is to tell the story of the Hanford Openness Workshops (Workshops) and extract lessons that can be applied to other DOE sites and to the current task of long-term stewardship. The Workshops were a series of nine meetings held from 1997 to 2000 to identify and address issues impeding the availability and understanding of information important to public health, the environment, and decision making at Hanford. They were a collaborative effort among the DOE-Richland Operations Office, Consortium for Risk Evaluation with Stakeholder Participation (CRESP), University of Washington, Oregon Office of Energy, Washington State Department of Ecology (WA Ecology), and regional tribal and public interest organizations. In this case study of stakeholder involvement in the nuclear waste cleanup context, we focus on both the process of the Workshops, as well as their products. The Workshops illustrate that stakeholders,[1] decision makers, technical specialists, and Tribes can put aside their differences, focus on common goals, and make constructive recommendations – even in a highly controversial environment. Moreover, the Workshops exemplify a real-world model of stakeholder involvement that provided critical lessons and recommendations for more open and transparent decision making at DOE facilities and show how DOE can move beyond a traditional "decide-announce-defend" mode of interaction with its stakeholders. Several Workshop coordinators and participants are themselves the authors of this paper, which represents a self-evaluation and record of how the Workshops worked and their outcomes. We welcome further independent scrutiny of the process and outcomes.

The remainder of this paper describes the process and outcomes of the Workshops in a case study format (Yin, 1993). After reviewing the connections between openness and long-term stewardship, we describe key events that helped in the site's transition from a highly secretive organization to one that is less so. Next, we describe the format of the Workshops, and its key products and findings. The fourth section explores factors critical to the Workshops' success, drawing from established principles of public involvement, and goals established by participants. Finally, we describe some of the key challenges faced by the Workshops, and explore conclusions and lessons for long-term stewardship. An epilogue provides an update on openness in the DOE context since the end of the Workshops.

Openness and Long-Term Stewardship

The study of openness is highly relevant to the task of long-term stewardship. Several efforts that have looked at stewardship of DOE legacy wastes have explicitly called for stakeholder involvement in the development of stewardship plans and policies (Applegate & Dycus, 1998; NRC, 2000, 2003; Probst & Lowe, 2000). Openness policies and practices of an agency establish context and tone for this involvement and thus are a critical element to understand and develop. Stewardship decisions must be guided by consideration of contextual factors that include values and preferences of interested and affected parties. In order to obtain an accurate understanding of values, an open and frank exchange of information with stakeholders in an ongoing two-way dialogue is needed.

Long-term stewardship also calls for strong accountability measures (Applegate & Dycus, 1998; NRC, 2000, 2003 ; Probst & Lowe, 2000). Openness is closely tied to accountability as well. Any major government activity should be open to the scrutiny of the tax-payers, who finance these activities (Buiter, 1999; Goldberg, 1994; Stiglitz, 1999). Nuclear waste cleanup and subsequent stewardship activities are thought to be the largest long-term government liability after the National Debt (US DOE, 1995a). The cleanup and long-term stewardship decision processes must be transparent to legislators, Tribes, the public, and citizens groups who represent them. In other words, those who are interested need to understand enough to evaluate activities and assure that tasks are completed.

Good records management is an additional requirement for long-term stewardship. Nuclear waste cleanup decision records need to be available and understandable to a wide range of individuals and organizations

(Applegate & Dycus, 1998; NRC, 2000,2003; Probst & Lowe, 2000). Records management review and maintenance processes should involve interested and affected parties to ensure that specialized knowledge needed for nuclear waste management is explicitly stated in the indexing systems, and not just implicitly understood by contractors and DOE employees.

THE ROAD FROM SECRECY TOWARD OPENNESS

Several major laws were passed in the U.S. between the mid-1960s and mid-1980s that promoted openness (e.g., Freedom of Information Act 1966), worker health and safety (e.g., Occupational Safety and Health Act, 1970; the Federal Hazard Communication Standard, 1972), and environmental protection (e.g., the Resource Conservation and Recovery Act, 1976; the Comprehensive Environmental Response, Compensation, and Liability Act, 1980; the Emergency Planning and Community Right to Know Act, 1986). These laws allowed citizens and watchdog groups to question the need for blanket secrecy at the former nuclear weapons facilities with increasing frequency and effectiveness. Documents were released, accounts of disturbing activities came to light, and additional demands for openness were made. Thus, Hanford started down a winding road toward openness (Table 1).

A key event occurred in 1984, when DOE-Richland invited the public and Washington State to comment on the environmental assessment for the Basalt Waste Isolation Project – one of three proposed sites for a national high-level radioactive waste repository. Soon after (1986), a Freedom of Information Act (FOIA) request from the Hanford Education Action League compelled DOE-Richland to release 19,000 pages of unclassified and once-classified information pertaining to historical operations (D'Antonio, 1993). Included in this material was an account of the 1949 "Green Run," in which 11,000 Ci of radioactive iodine and 16,000 Ci of xenon were intentionally released from the Hanford Site (Robkin, 1995).[2] Alarmed at what the government might be concealing, the public demanded greater access to documents (D'Antonio, 1993).

In 1986–1987 DOE-Richland sought public comment on the Hanford Defense Waste Environmental Impact Statement (US DOE-Richland, 1986), which analyzed options for treating and disposing high-level tank wastes. In 1989, DOE signed the Hanford Federal Facility Agreement and Consent Order (also called the Tri-Party Agreement or TPA) (TPA, 2003). This agreement committed DOE-Richland to clean up Hanford with the oversight of the Washington State Department of Ecology and the US

Table 1. Hanford's Road to Openness (1970s–2000).

Late 1970s–1984	DOE names the Basalt Waste Isolation Project at Hanford as one of three candidate sites for a national high-level radioactive waste repository. In 1984, DOE invites the public to participate in the environmental assessment. It is the first significant public process at Hanford and the first opportunity for the State of Washington to play a larger role in deciding Hanford's future
1986	DOE-Richland releases 19,000 pages of documents outlining Hanford's historical operations and releases of contamination to the environment. For the first time, the public gets significant information about Hanford's production history and related contamination. Public interest in Hanford's skyrockets
1986–1987	DOE conducts its first public process to support Hanford decision-making. The Hanford Defense Waste Environmental Impact Statement (US DOE-Richland, 1986) analyzed options for treating and disposing some Hanford wastes
1987–1995	The Technical Steering Panel (TSP), an independent advisory group, reviews progress on attempts to estimate public radiation doses from Hanford releases (the Hanford Dose Reconstruction Project). Several TSP participants are given clearances allowing them to view still classified documents
1989	The Hanford Federal Facility Agreement and Consent Order, the "Tri-Party Agreement" (or TPA) is signed by Washington State Department of Ecology, the U.S. EPA, and DOE-Richland, requiring the Hanford site to abide by national and state environmental laws for the first time (TPA, 2003)
1990	"Community Relations Plan" for the Tri-Party Agreement is signed (TPA, 2002)
1992	The State of Washington produces a report recommending an open process to resolve employee concerns (Brock, 1992). Ultimately, the Hanford Joint Council on Resolving Significant Employee Concerns is created, giving those outside Hanford an opportunity to review and aid in the resolution of employee concerns cases
1994	Hanford Advisory Board meets for the first time. See http://www.hanford.gov/boards/hab
1994	The Hanford Summit II meeting highlights the issue of openness and its importance for DOE to accomplish its cleanup goals. A Hanford Openness Panel is proposed
1995–1997	Public interest organizations advocate for an ongoing openness panel. After two years, negotiations with DOE result in the creation of Hanford Openness Workshops

Table 1. *(Continued)*

May, 1997	The first Hanford Openness Workshop is held
August, 1998	The first major Workshop report is issued (HOW, 1998c)
December, 1998	A response to the first Workshop report is received from DOE-Richland (US DOE-Richland, 1998)
February, 1999	The second series of Hanford Openness Workshops begins
September, 1999	DOE-Headquarters responds to the first Workshop report (Huntoon, 1999)
August, 1999	DOE-Richland announces that nearly 2 million pages of documents have been de-classified and that all documents are scheduled for de-classification by the end of Fiscal Year 2003
December, 1999	The Workshop issues its second major report (HOW, 1999e)
February, 2000	Final Hanford Openness Workshop held, featuring face-to-face discussion between Workshop participants and several directors of DOE-Richland

Environmental Protection Agency (EPA). Prior to the TPA, Hanford, like many government facilities, had not been required to comply with state and federal environmental laws. The Hanford TPA was the first agreement of its kind for a major government facility and paved the way for similar agreements at other DOE and Department of Defense sites. The TPA contained explicit requirements for public involvement in the "Community Relations Plan" (TPA, 2002), first signed in 1990.

Early efforts to commence public involvement processes and the release of previously classified documents in 1986 together helped expand public interest in Hanford. People wanted answers to both basic and detailed questions, such as: Have people living downwind of Hanford received harmful doses of radiation? Where are the waste sites located? What are the risks posed by wastes currently at Hanford? What waste sites should receive top priority in the cleanup? What is in the waste and how much is there? What technologies may be applicable for cleaning it up?

Answering such questions requires an integration of sound science with extensive public participation. This is a challenging task for any organization, especially for DOE-Richland because of its history of secrecy and other factors. Until the Cold War ended, Hanford's primary mission was production of nuclear materials, not waste management. Thus, historical records were kept for the purpose of understanding weapons material production, not to facilitate cleanup, or public understanding of cleanup. And, many records that did contain relevant information had been classified. Although many records were "available" to the public in reading rooms around Washington State after 1986, it was difficult to know what was in them.

For example, many titles do not reflect the full contents of certain documents. Partial censorship of documents only served to escalate these issues. Tribal Nations interested in certain kinds of data were not able to tell simply by looking at a title if there was relevant material in a document (HOW, 1999d). In addition, titles were also full of jargon that was (and still is) difficult to decipher. One source of jargon comes from the complex alphanumeric scheme used to label Hanford's numerous geographic subdivisions. For example, all the reactors (9 in total) have different letter designations (B, C, D, DR, F, H, KE, KW and N), and they are all located in the "100 Area" which lies adjacent to the Columbia River (Fig. 2). For purposes of cleanup,

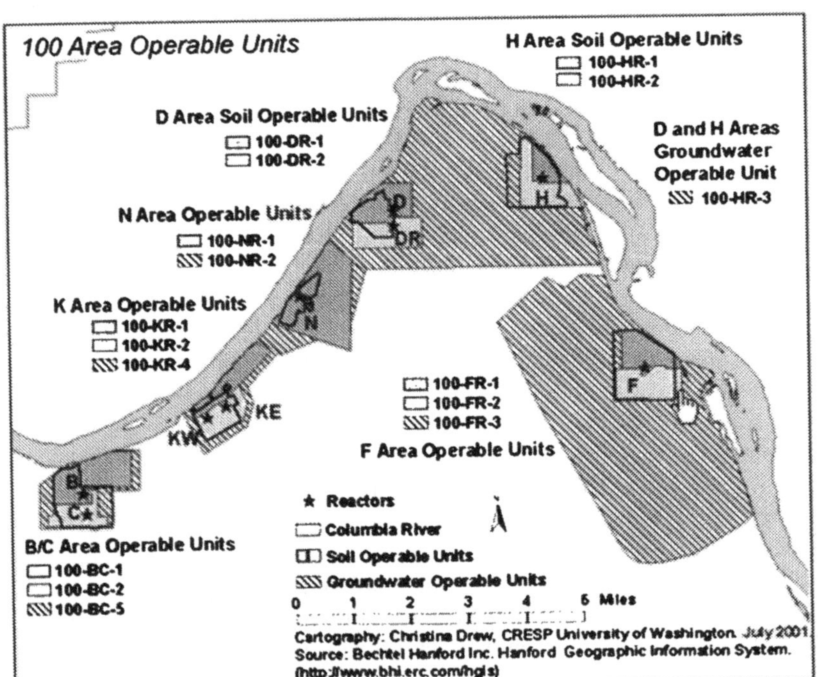

Fig. 2. Hanford 100 Area Operable Units.

administrative boundaries, called "Operable Units," were established based on the original system. Thus today it is common to discuss the "100-K Area," the "100-BC-1 Operable Unit" or the "116-B-14 waste site". Over 50 Operable Units and 2000 individual waste sites have been identified within Hanford's nearly 600 square miles. Thus, searching databases without understanding this nomenclature is extremely challenging.

Several issues that arose during the current era at Hanford have decreased the credibility of DOE-Richland, its contractors and DOE-Headquarters. Despite citizen concerns, DOE maintained that Hanford's high-level radioactive tank waste would not contaminate groundwater. Although DOE acknowledged that past operations during the 1950s and 1960s had contaminated the groundwater, they claimed that similar waste from underground storage tanks would never reach the groundwater because the soil characteristics would hold the waste in place until radioactivity had decayed. DOE maintained this position even though internal documents (not released to the public) spoke of direct risks to groundwater (Hanford Education Action League (HEAL), 1992). When new technology was applied to the problem in the late 1990s, DOE discovered that waste from the tanks had already reached the groundwater (Hanford Reach, 1997). Hence, an advisory committee to the Secretary of Energy noted that, "(T)here is widespread lack of trust in DOE's radioactive waste management activities. That distrust is not irrational, nor can it be discounted merely as a manifestation of the 'not-in-my-back-yard' syndrome" (SEAB, 1993, p. v). This basic lack of trust provides the backdrop for all interactions between DOE and its stakeholders (including regulators and Tribes).

In the early 1990s, the then-Secretary of Energy Hazel O'Leary committed to developing a culture of openness and straight talk at DOE. Throughout her term, she implemented and expanded openness initiatives by holding press conferences and releasing vast quantities of previously secret information and commissioning reports through her advisory board. O'Leary established the Office of Human Radiation Experiments (http://tis.eh.doe.gov/ohre/) that produced a review of all such experiments (US DOE, 1995c). The Secretary of Energy Advisory Board (SEAB) also emphasized the importance of openness in its 1993 report stating that democracy "demands that public institutions operate in a transparent manner, that they adopt processes that not only permit but encourage broad segments of the population to participate, and that no segment finds itself permanently a 'loser' in policy controversies" (SEAB, 1993, p. v).

Several National Academy of Sciences reports released in the mid-1990s also focused on improving DOE cleanup (NRC, 1994, 1995). These studies called for increasing public and tribal government participation in decision-making processes, especially for budgeting and priority setting. The studies also highlighted the need to integrate scientific and technical information and communicate it to stakeholders, and to recognize that cleanup decisions needed to include discussion about values, not just facts and technologies.

In 1994, at a meeting called the Hanford Summit II, DOE-Richland proposed to convene a Hanford openness panel. The panel's mission would be to increase public access to documents, transparent decision-making, accountability and employee openness, and to eliminate reprisals for workers who raised concerns. Stakeholders, tribal nations, and regulators aggressively pursued this commitment to Hanford openness, but more than two years passed before a format acceptable to DOE was found. Finally, agreement was reached in 1997 and the Workshops were formed. The following section describes the process and major accomplishments of these Workshops.

THE HANFORD OPENNESS WORKSHOPS

An initial series of Workshops were held between 1997 and 1998 and a second series was held from 1999 to 2000. Their mission was to resolve issues impeding the availability of information important to public health, the environment, understanding, and decision-making at Hanford (HOW, 1997a). They were convened to assist and inform DOE-Richland on issues related to (1) declassification, (2) improving public access to Hanford information, (3) government and contractor accountability, (4) creating an open and transparent decision-making process, and (5) institutionalizing openness throughout DOE-Richland and its contractors' activities. The Oregon Office of Energy (OOE), the Washington State Department of Ecology (WA Ecology), the Consortium for Risk Evaluation with Stakeholder Participation (CRESP), the University of Washington, and DOE-Richland developed the Charter and implementation plan for these Workshops. This coordinating group worked collaboratively with regional citizen and tribal representatives.

Funding for the Workshops was provided by DOE-Richland and WA Ecology through grants to the University of Washington. At the time, many faculty and staff at the University of Washington were working with CRESP,[3] with a focus on Hanford cleanup. These individuals were involved in facilitation of the Workshops, and provided technical and administrative staff support to the Workshops. Their role in the Workshops was one of CRESP's efforts to improve the transparency of decision-making at Hanford and improve communication between stakeholders, DOE, and Tribes. Several Workshop participants later contributed to a related effort by CRESP researchers to understand the meaning of "transparency", and implement a more transparent public record at Hanford (Drew, 2002; Drew & Nyerges, 2004; Drew, Nyerges, & Leschine, 2004). CRESP/University of Washington researchers and staff coordinated meeting logistics, provided technical consultation to workshop participants (e.g., information on archiving documents, assessment of data quality, and data accessing techniques), and prepared fact sheets and reports based on the participants' conclusions and recommendations. CRESP provided a professional facilitator, administrative coordination, and technical assistance to the Workshops.

Workshop Participants

Workshop participants were selected by a membership committee comprised of representatives from the coordinating organizations (DOE-Richland, WA Ecology, OOE, CRESP, and the University of Washington). The intent was to involve the perspectives of a wide variety of stakeholders, government agencies, and tribal nations. No legal requirement specifically mandated that the Workshops be conducted or that any individual or interested party participate. Therefore, all participants were involved solely because of the value they placed on the subject of openness at DOE. The Workshop's Charter required a varied membership including representatives from three public interest organizations and three tribal nations, (one each from the Umatilla, Yakama and Nez Perce Tribes), representatives from a "Whistleblower" advocate, the States of Oregon and Washington, the "Downwinder" community (those most likely to be affected by atmospheric releases of radionuclides from Hanford), a Columbia River protection group, and academia, as well as an ad hoc member (e.g., journalist, historian, League of Women Voters, teacher), and a current or former Hanford employee with relevant recent experience (Table 2).

Workshop Process

The Workshops established procedural guidelines that were accepted by all participants. The Charter (HOW, 1997a), initially developed by the coordinating organizations, was subsequently revised and adopted by all

Table 2. Organization Participating in the Hanford Openness Workshop.

Workshop Participants	Workshop Technical Specialists and Agency Contacts
Columbia River United	City of Seattle, Office of Technology
Confederated Tribes of The Umatilla Indian Reservation	Consortium for Risk Evaluation with Stakeholder Participation
Government Accountability Project	Fluor Daniels Hanford
Hanford Downwinders Coalition	Hanford Health Information Archives
Hanford Workers (one former and one current)	JK Research Associates
Heart of America Northwest	National Archives and Records Administration
Nez Perce Tribe	Oregon Office of Energy
Oregon League of Women Voters	Pacific Northwest National Laboratory
Oregon Office of Energy	University of Washington
Physicians for Social Responsibility	Human Subjects Division
University of Washington, Evans School of Public Affairs	Department of Environmental Health
US Dept. of Energy, Richland Operations Office	Department of Geography
Washington Dept. of Ecology	U.S. Dept. of Energy, Richland Operations Office
Yakima Indian Nation	Office of River Protection
	Office of Chief Council
	Office of Declassification
	External Affairs
	Public Involvement
	Environment, Safety and Health
	Indian Nations Program
	Employee Concerns Program
	U.S. Department of Energy, Headquarters
	Office of the Assistant Secretary for Environmental Management
	Office of Declassification
	Yakima Indian Nation, Office of Environmental Restoration and Waste Management

participants. It provided the mission, goals, and purpose of the Workshops, describing the scope of issues for discussion, meeting logistics, membership, funding considerations, public involvement, and press inquiries. The Workshops also officially adopted a set of ground rules (HOW, 1997b), which spelled out behavioral expectations and committed Workshop participants to interacting with one another in a fair and respectful manner. Participants frequently referred to these ground rules during Workshop meetings and several stated that these ground rules were essential for facilitating constructive dialogue and debate.

Consensus-Based Decision Process

The Workshops employed a variety of mechanisms to provide voice and authority to each participant without allowing a disproportionate accumulation of power in any single group or individual. One approach was the consensus-based decision process adopted in the Charter, chosen to prevent marginalization of participants who were not of the majority's opinion. Workshops organizers felt that operating by consensus would encourage participants to focus discussion on areas where there was broad agreement, enabling the development of stronger and well-supported recommendations. All major recommendations and issues were put before all participants and were not considered "approved" until each had the opportunity to comment, suggest revisions, and object if they were not in agreement with the recommendation. When differences of opinion arose, the participants attempted to reach common ground through facilitated discussion. If differences of opinion could not be resolved, the Charter permitted a majority recommendation to be forwarded with attached alternative views. However, this never occurred because participants chose to focus on recommendations that had strong and broad support. The Charter required DOE-Richland to respond to written recommendations from the Workshops prior to taking action on the subject of a recommendation.

Professional Facilitation/Spokesperson

CRESP provided a professional facilitator for the workshops, as required in the Charter. Two participants were elected to serve as spokesperson and co-spokesperson, representing the Workshops to the public and the media. Participants believed that electing a chairperson would create an imbalance among themselves and endanger their ability to reach consensus. Instead they requested that the facilitator also perform the role of chair, i.e., lead the meetings, balance the agendas, ensure equal participation, etc.

Open Meetings

The Charter required that all Workshops be open to the public and conducted within the parameters of Washington State's Open Public Meetings Act and the University of Washington's "Community-Based Research Principles" (University of Washington, 2004). The public was afforded an opportunity to comment at all meetings.

Major Workshop Achievements

The most important achievement of the Workshops was a new definition of openness. Prior to the Workshops, Secretary O'Leary defined openness as: "open, ongoing, two-way communication, both formal and informal, between the Department of Energy and its stakeholders" (O'Leary, 1994). The initial effort was focused on declassification, but participants soon agreed that declassification was only one component of openness. Ultimately, participants expanded the definition to include the whole structure of Departmental interactions, in which access to information enables individuals and groups to contribute meaningfully to decision-making processes:

> Openness is when DOE works with stakeholders to find answers for problems that impede access to information and creates decisions that are reflective of stakeholder needs. Openness is when all pertinent information is available for public review and understandable prior to decision-making. Openness is an open and transparent decision-making process. Policy changes are the first steps in bringing about change, but only the first steps. Openness is successful only when the policies are institutionalized, creating a culture of openness. (HOW, 1998c).

The Workshop process itself provided a model for addressing openness issues at other DOE sites and other public participation processes. It also brought clarity to a wide variety of issues that are closely linked to openness and developed methods and approaches to address those issues. At the first Workshop, participants established a number of smaller working groups to address specific topics related to openness concerns, including: historical documents, performance measures, employee concerns, document title review, information tools, declassification, public involvement, and tribal concerns. Each group worked both during and between workshops to investigate issues of interest, develop fact sheets, and form recommendations on their specific issue areas. Participants discussed ways to institutionalize their ideas (i.e., create a culture of openness) and to put them into practice in the DOE context. This resulted in a significant body of work: two reports outlining recommendations for implementing openness at both Hanford and at the DOE headquarters level (HOW, 1998c, 1999e), eight short (2–4 pages each) fact sheets on a range of issues, (access to documents, background on the Workshops, classified information, document review, Internet resources, openness and security, public involvement, and tribal openness concerns), and several letters from the Workshops to decision makers on openness-related topics (Frei, 1997; HOW, 1997c, 1999e, 1999f, 1999g; Huntoon, 1999; Wagoner, 1998). All meeting materials and

Table 3. Key Workshops Recommendation to DOE (Abbreviated list) (HOW, 1998c).

Continue the Workshops to build trust, identify and clarify priorities for declassification, ensure openness is a long-term commitment, resolve problems, and support continued funding for openness, participation and declassification

Include in contracts incentives for achieving openness goals and penalties for not achieving these goals. Also include incentives and penalties for compliance with treaties and other federal obligations to tribes and tribal nations

Use the Workshops' draft evaluation plan and positive/negative examples to develop more useful and comprehensive public involvement evaluation mechanisms

Suspend or revoke contracts if a contractor has engaged in "whistleblower" reprisals or maintained a hostile environment to employees reporting concerns

Provide ongoing training for supervisors and workers. Hold managers accountable for reprisals. Provide new employee orientation on openness

Reverse current DOE policy of reimbursing contractors for litigation costs.

Simplify employee concerns processes, paying attention to their "trackability"

To aid future declassification, *develop* policies and marking systems that identify the sensitive information in newly created documents and the laws that apply

Ensure declassification and openness policies apply to all contractors and Hanford-related records, including those held, administered or controlled by contractors

Include a plan for ultimate public release in all DOE document reviews, including those conducted as part of legal processes

Increase the number of documents and finding aids (such as Internet accessible search engines and methods for establishing keywords and indexes) electronically available. All released records should be electronically searchable and available via the Internet

Institute a mechanism by which the public can conduct full text searches on document abstracts. Abstracts reveal more about a document's content than do titles

Develop and implement systems to perform "data mining" – looking for patterns of information relevant to a variety of searches – on DOE documents

Recognize and implement the government-to-government relationship between the United States and the tribes reflected in the DOE American Indian Policy

Convene a meeting with senior managers and the Workshops to discuss and strategize how to achieve the goals delineated in the Workshops reports

publications were made available on an Internet site (HOW, 1998b). Table 3 provides an abbreviated list of HOW recommendations, and Table 4 summarizes responses from DOE-Richland.

CRITICAL SUCCESS FACTORS

Participants in the Hanford Openness Workshops widely regarded them as a success, based on the initial goals set out for the interaction, as well as several key principles that have been highlighted in public involvement and

Table 4. Examples of DOE-Richland's Response to the Workshops (Simplified for the Sake of Brevity). (HOW, 1998c; US DOE-Richland, 1998).

In the response, DOE-Richland agreed to ...
... continue funding the Workshops for a new set of workshops (the 1999 series)
... the Workshops' comprehensive definition of "adequate notice" for public meetings and other public involvement activities
... implement new methodologies for a "safety conscious" workplace
... continue support for a council designed to resolve complaints of workers
... establish a monitoring program for employee trust and confidence issues
... increase the availability of health and environmental information to support studies
... issue an annual policy statement regarding employee climate issues and concerns
... begin development of a system identifying the types of information typically found in classes or types of documents (e.g., location of radioactive or hazardous materials, releases to the environment, and exposure of site employees)
... continue full funding of efforts to declassify documents
... make the Hanford Information Systems Index (an index of Hanford databases) public
... enhance the user-friendliness of the Declassified Document Retrieval System
... conduct a workshop devoted to Tribal openness concerns
DOE disagreed with seven recommendations, declining to ...
... include contract incentives of five or six percent for contractors achieving openness goals. DOE has a subjective evaluation on openness issues which it felt was adequate
... suspend or revoke contracts if a contractor has engaged in whistleblower reprisals. DOE countered that such suspensions would impact cleanup requirements to an unacceptable degree
... institute a policy to make individual managers accountable for whistleblower reprisals although DOE committed to giving the recommendation further consideration
... blanket release all records with a classification below "Confidential." According to DOE, Administrative Orders prevent the agency from blanket release of such documents
... identify newly generated documents of records which require protection under the Privacy Act
... include incentives in contractor evaluations for openness to address compliance with treaties and other federal obligations to the Tribes. DOE holds that it is responsible for compliance with federal obligations to Tribes and cannot delegate the compliance responsibility to its contractors

community-based participatory research literature (Arnstein, 1969; Beierle, 2002; Bradbury, Branch, & Malone, 2003; Drew et al., 2003; IAP2, 1999; Israel et al., 2001; Kinney & Leschine, 2002; Kleinman, 2000; NRC, 1996; US EPA, 2003). These principles include attention to process (i.e., getting the participation right), inclusive representation (i.e., getting the right participation), commitment to openness, technical assistance, providing a clear value to the agency, and accountability. This section discusses these principles and illustrates their importance to the Workshops using examples from the fact sheets and reports, and reviews these achievements in light of the original aims set out in the workshops Charter.

Attention To Process

Many research articles about public participation activities recommend paying specific attention to the interaction process (Bradbury et al., 2003; Drew et al., 2004; IAP2, 1999; Israel et al., 2001; Kinney & Leschine, 2002; Renn, Webler, & Wiedemann, 1995; US EPA, 2003). The National Research Council (NRC) refers to a need for "getting the participation right," which means that the process is responsive to participants' needs, that concerns are addressed, and that participation has shaped the way problems are defined and understood (NRC, 1996). Participants viewed operational guidelines, established in the Charter and ground rules, as especially critical to the success of the Workshops. In particular, having a facilitator to set agendas, act as arbiter of disputes, keep discussion on track, and remind participants about ground rules led participants to produce consensus-based recommendations, reports, and fact sheets.

Inclusive Representation

The NRC also points to the need for "getting the right participation" (NRC, 1996). That is, the appropriate range of individuals and organizations participate to foster credibility and achieve fairness. In the Hanford context, there are a wide variety of interests (Boiko et al., 1996) so a broadly representative range of views about Hanford cleanup is needed. Diversity was required by the Workshops Charter and as a result diverse perspectives were represented in the Workshops (see Table 2). The example below illustrates how the Workshops were able to bring an innovative perspective to Tribal concerns through direct participation of Tribal representatives.

Example: Tribal Openness Concern

In the U.S., Tribes are recognized as sovereign entities that have rights to a full government-to-government relationship with the U.S. Government and its agencies (Clinton, 1998; d'Errico, 1998; HONOR, 1998; Jim, 1994). This relationship is challenging for bureaucracies such as DOE to honor. The

Openness Workshops participants hosted a Tribal Openness Workshop in June of 1999 to discuss the unique openness concerns of Tribes.

At the Tribal Openness Workshop, tribal representatives emphasized that DOE must make a stronger effort to acknowledge, understand, and respect the cultural values and communication methods of Tribes. Specific areas of concern highlighted during the round table discussions included that:

- Government-to-government consultation requires dialogue between DOE and each Tribe; providing documents is not enough.
- Tribes are sovereign governments; therefore, processes that involve Tribes alongside "the public" honor neither government-to-government relations nor legal obligations.
- DOE must recognize the distinctness of each Tribe and respect intertribal differences.
- DOE must respect tribal methods of communication and decision-making. Many tribal members – particularly elders – are wary of written or recorded communication.
- DOE must respect tribal desires to keep certain cultural information confidential (e.g., locations of burial grounds, sites of spiritual significance, gathering sites, other cultural resources).
- Declassification is just one part of tribal openness concerns. The process of making information available and accessible is as, if not more, important.
- Declassification deals with openness about past activities. Present-day openness is also important to the Tribes.

These workshops provided the DOE classification team the opportunity to discuss openness issues directly with tribal representatives. Meaningful dialogue among DOE and tribal representatives allowed them to come closer to a shared understanding of each other's concerns. The tribal representatives described how information that DOE considers routine can be interpreted very differently. This resulted in a significant change in the DOE team's thinking. They went from not thinking at all about tribal information needs to working directly with Tribes to develop keywords and indexing to create declassification procedures that are more sensitive to tribal concerns. The Workshops developed a fact sheet on Tribal Openness Concerns (HOW, 1999d), which highlighted the Tribe's lack of resources and information access issues, emphasized the need for continued dialogue among DOE and Tribes, and provided a list of relevant resources (Clinton, 1998; d'Errico, 1998; HONOR, 1998; Jim, 1994).

Commitment of DOE, Regulators and Individuals

Broad-based agency and stakeholder commitment is essential for effective public involvement and openness activities (IAP2, 1999; Kinney & Leschine, 2002; US DOE-Richland, 1997; US EPA, 2003). In part reacting to earlier, less-than-successful stakeholder involvement activities (e.g., the Columbia River Comprehensive Impact Assessment – see Kinney & Leschine, 2002), workshops organizers felt strongly that DOE's commitment over several years was critical. The Charter thus laid out that DOE would fund the workshops over a two-year period, respond in writing to all written input from the Workshops, and provide staff support (HOW, 1997a). DOE not only met these requirements, but also encouraged personnel from at least nine different internal programs or offices to interact with participants. Also, the top managers at Hanford, representing the Richland Operations Office and the Office of River Protection, personally participated in the final Workshop in 2000.

DOE's funding support was matched by the collaborating organizations'. Thus, the workshop series, valued at approximately $100,000, was conducted at a hard cost to DOE of only $10,000. In addition, all individuals who chose to participate approached the Workshops with energy, creativity, and enthusiasm. Specifically, the commitment and dedication of several DOE-Richland staff members was critical, suggesting that one way to change the culture of a large bureaucracy is a single employee at a time. This commitment by all involved played an important if intangible role in the Workshops' success.

Clear Value to Agency

During the Workshops, DOE-Richland interacted directly with participants and became better informed about their concerns and positions, while participants likewise became more involved in DOE's plans and gained a better understanding of DOE-Richland's internal processes. As a result, participants were better able to identify why particular concerns were not resolved and how to work towards resolution. Such a focus on two-way communication – rather than the one-way communication typified by the "decide-announce-defend" model of decision-making – allowed better identification and resolution of issues and disputes. As a whole, the Workshops provided a direct value to DOE-Richland because they produced a sustained, iterative dialogue about Openness with tribal, stakeholder, and regulator communities. Specific examples of direct benefits included focusing on employee concerns and developing "success stories."

Example: Sustained iterative dialogue with DOE-Richland

The International Association for Public Participation (IAP2) has highlighted the need "for direct advice and innovation in formulating solutions" (IAP2, 2000). The Workshops illustrated a clear value to DOE-Richland in this area by providing a model for direct, iterative communication among DOE-Richland decision makers, stakeholders, and Tribes. Often, the route of statements made by stakeholders and Tribes to a decision maker's ear and back is circuitous and results in inaccurate message delivery. By focusing on direct tribal/stakeholder/decision maker communication, the Workshops fostered better understanding and productive discussion (HOW, 1999a, 1999b). Prior to the Workshops, DOE received only general calls for "more openness" from stakeholders. Lacking specifics meaningful responses were difficult to develop. Participants' directed and sustained focus on specific issues allowed discussion of approaches and solutions to help resolve challenges.

The Workshops' activities were also discussed at the national level. In the first Openness Report 33 recommendations were directed at DOE-Headquarters (HOW, 1998c). A response from DOE-Headquarters to these recommendations was received in 1999 (Huntoon, 1999) and the Workshops began a dialogue with DOE-Headquarters on those recommendations. Recommendations were also shared with Site Specific Advisory Boards for other major DOE facilities, since Hanford was the only site with an openness initiative. The Workshops developed a relationship with the Secretary of Energy Advisory Board's Openness Advisory Panel (OAP), a group that was established to advise the Department about the status and strategic direction for its classification and declassification policies across the nation. The first "field meeting" of the OAP was held near Richland, WA in February 1998, in part because of the major contributions of the Workshops (Openness Advisory Panel, 1998).

Example: Employee Concerns

Hanford has a history of "whistleblowers," who, characteristically, are employees that raise safety concerns and are then fired or harassed on the job (Lynch, 1999). Often, whistleblower cases end up in court when an employee sues the contractor. Given DOE's policy to reimburse contractors for litigation costs (US GAO, 2003), the participants argued that contractors have an incentive to fight any and all cases vigorously because their corporate resources are not at risk. The Workshops recommended that DOE-Richland reverse the policy of reimbursing contractors for these litigation costs, as well as revoke the contracts of any contractors who engage in whistleblower reprisals. Participants also investigated the employee concerns resolution processes, but could not ascertain how employee concerns are prioritized at each step of consideration, what percentage of employee concerns are resolved versus unresolved or what follow-up activities are taken when issues are unresolved. The Workshops recommended that employee concerns programs be simplified, with the goal of increasing accountability, transparency, and "trackability" of the concerns resolution processes, and that managers receive better training in these areas. The participants also identified several examples of information sources that describe and understand how other organizations have approached methods to institutionalize a "safety culture" (Behrens & Muller, 1993; Birkner & Birkner, 1996a, 1996b, 1996c; Jorgensen, 1991; McSween, 1995; Reason, 1995), and how these can be applied in the DOE context. The Workshops' advice about employee concerns provided a clear value to DOE, who agreed to continue support for a council designed to resolve complaints of workers, and establish a monitoring program for employee trust and confidence issues (US DOE-Richland, 1998).

Example: Success stories

Equally important to the Workshops' achievements was the focus on success stories, not just on concerns and recommendations. Throughout the Workshops, Participants raised many issues, but always discussed them with the aim of developing constructive recommendations for DOE-Richland. A suggestion to include success stories originated with a DOE-Richland manager after the first series of Workshops. Participants recognized that DOE-Richland had indeed made some important positive changes that deserved recognition and reinforcement. Examples of such successes were included in the Progress Report to give credit to DOE (HOW, 1999e). Including both positive and negative examples in the Progress Report – and explaining why they were viewed as positive or negative – allowed a fuller understanding about where openness was working and where it was not. An example of a success story identified in the area of public involvement was the Columbia River Comprehensive Impact Assessment Meetings (1996), which were held in four locations around the Northwest. The meetings resulted in strong support for the impact assessment, and were viewed as successful because the assessment team and the agencies cooperated about notifying the public, preparing information, and setting the format for the meetings.

Participants identified the Tank Waste Environmental Impact Statement Meetings (1996) as a negative example. The written material for these meetings was full of acronyms and jargon. Not enough advance notice was given. Organizers did not involve Oregon or interest groups in arranging the meeting. DOE-Richland used a very expensive facility. Contractors refused to switch to an informal mode, even after being asked directly and politely to do so by the public. The overheads were difficult to read, contained jargon and acronyms, and altogether too much material. More bureaucrats attended the meeting than public. DOE-Richland was asked by the public to stop using the overheads and begin a dialogue, but refused. As a result, some public attendees left.

Technical Assistance

The Openness Workshops made use of many forms of technical expertise, which have long been seen as a critical component of stakeholder interactions (Arnstein, 1969; Drew et al., 2003; Kleinman, 2000; US EPA, 2003). One form of expertise came from participants themselves. The collective experience of Workshop participants included a strong command of the issues, access to additional technical resources, and highly organized constituencies. Furthermore, while there has been great turnover among managers at DOE and its contractors in the past 10–15 years, many of the individual stakeholders and tribal representatives who were involved with the first release of Hanford information took place in 1986 are still active. When additional expertise was needed, experts from various organizations from DOE-Richland and DOE-Headquarters, contractors (including Pacific Northwest National Laboratory and Fluor Daniels Hanford), the University of Washington, and others were invited to participate in discussions (see Table 2). All of the working groups benefited from technical expertise. Declassification and information tools are discussed below as examples.

Example: Declassification

At times, DOE-Richland is required to review, declassify, and release to litigants documentation for court cases. Participants saw this as a significant opportunity for openness and recommended that DOE-Richland develop plans for public release of documentation reviewed as part of legal processes and as a part of the day-to-day business of the Department. The Declassification working group brought together experts from DOE-Richland's declassification team, as well as experts from the University of Washington,

DOE-Richland Indian Nations Program, DOE-HQ Office of Declassification, Hanford Health Information Archives, and the Pacific Northwest National Laboratory. Recommendations and support from the Workshops led to the implementation of an online Hanford Declassified Document Retrieval System (US DOE-Richland, 2004), which by 2004 included nearly 120,000 declassified documents and 77,000 declassified photographs related to Hanford's operations and cleanup activities. The workshops also supported DOE's nation-wide declassification effort, the Open-Net database (US DOE, 2000).

Example: Information Tools
The Information Tools Working Group advised DOE-Richland about conducting key-word searches on documentation to determine the utility of titles as the basis for information searches. They also explored current technologies to improve information availability and accessibility by identifying and prioritizing automated information access tools. The Working Group investigated a number of techniques to review and extract meaning from massive quantities of data, such as optical character recognition (OCR) and "data mining" software like the SPIRE/STARLIGHT software developed by the U.S. Government. Unfortunately, many scanned images are not of adequate quality to be used by data mining software. Participants also recommended that future declassification would be aided by developing policies and marking systems that identify sensitive information in newly created documents. The group heard from experts from the City of Seattle Office of Technology, the University of Washington Evans School of Public Affairs, the DOE-Richland, and DOE-HQ Declassification offices, the Pacific Northwest National Laboratory, and the National Archives and Records Administration. Face-to-face interaction with these experts was essential for developing Workshops recommendations.

Accountability

Accountability is another important element of public involvement (Arnstein, 1969; US DOE, 1995b; US EPA, 2003). Individually, Workshop participants were accountable to a community-based organization, constituency, or public agency. Collectively, the Workshops were accountable to the Hanford Advisory Board (HAB), a volunteer Board of over 30 stakeholder interests that advises the Tri-Party Agreement agencies on Hanford cleanup issues. Many of the members of the workshops were (and still are) also members of the HAB. The HAB monitors openness issues through its Public Involvement Committee and formally supported large sections of the Workshops' work (HAB, 1995, 1998). The HAB was instrumental in communicating the Workshops' recommendations throughout DOE's national Advisory Board community. The HAB's interest in openness continues to provide an extra accountability mechanism for both the Workshops and DOE-Richland. Specific measures to improve accountability also emerged from the Workshops in the areas of public involvement and contract performance measures.

Example: Public Involvement
The Workshops addressed many information access and stakeholder and public involvement issues in its recommendations, one of which was implementing and evaluating DOE-Richland's public involvement processes. Finding that a comprehensive evaluation mechanism was needed, participants developed a draft evaluation plan and recommended its use to DOE-Richland. The plan emphasized the importance of collecting meaningful information, but not in a way so detailed or time consuming that meeting participants are unwilling to respond. A multi-level strategy was recommended, including a simple 3 × 5 report card for the general public to complete at public meetings, a more detailed response form for the highly involved stakeholder community, and an annual evaluation to review comments, look for progress, and report to the public, Tribes and DOE. Participants suggested that a periodic (i.e., quarterly) summary of comments should be posted on the Internet, that web-based evaluation forms could be used to supplement the paper forms, and that an independent group (i.e., some one other than DOE-Richland) should design, sponsor, and collect forms. The working group also developed a series of interconnected values, goals, and performance criteria for evaluating public involvement (see Table 5) (HOW, 1999e, Appendix 14).

Example: Performance Measures
The performance measures working group focused on the importance of formally incorporating openness into DOE-Richland's contracting approach through measurable, contractual mechanisms. Contracts offer DOE an excellent opportunity to introduce performance measures for openness, including financial incentives and penalties for meeting or failing to meet openness targets. In the first report, participants recommended including contract incentives for achieving openness goals and penalties for not achieving these goals (HOW, 1998c). Recommendations included the

Table 5. Value Tree for Public Involvement Activities: Selected Values and Goals.

Value: Productive and Effective Public Involvement Activity: Pre-activity actions
Goal: Flexible style and format for meetings (design, layout, language)
Goal: Determine goals of public involvement activity prior to the activity (e.g., to inform the public, to get input on a process or decision, or to get feedback on a particular document). Work with states and stakeholders to decide what P.I. activity will meet goals
Value: Use creative and innovative ways to get information to the public
Goal: Effective advertising process (links to more targeted audience and is less costly).
Goal: notification strategy
Goal: Timing of public involvement
Goal: Provide feedback and evidence of how comments applied
Goal: Effective speakers
Goal: Provide appropriate meeting materials
Goal: Effective leadership
Goal: Effective moderators and facilitators
Value: Productive and effective public involvement activity: Post Activity
Goal: Timely feedback after meetings
Value: Cost effective public involvement activity
Goal: Use efficient advertising
Goal: Minimize number of agency staff and contractors
Goal: Work with local individuals to secure low cost meeting rooms
Goal: Maintain cost effective public involvement budget
Goal: Good attendance

Source: Adapted from HOW (1999e), Appendix 14.

following contractual obligations: specific performance measures for openness; both incentives for achieving and penalties for not achieving openness; an environment in which decisions made without disclosure are subject to reversal; traceable measures for citizen involvement in decision making; and independent mechanisms for review of compliance with openness objectives. Workshop participants agreed that contractors must be required to implement openness within their existing fee structure, and that contractors should not be allowed to increase the cost of negotiated contracts to include openness mechanisms.

Other metrics of success that the participants identified during the workshops included regular participation and attendance, production of reports and fact sheets, and the acceptance of recommendations by DOE. Meeting attendance remained high throughout the nine workshops (HOW, 1998b). Reports and fact sheets were produced and are still available online (HOW, 1998b). DOE responded in writing to the reports and did implement some of the recommendations, e.g., the list of keyword search criteria that was adopted by the DOE declassification team came directly from the workshops. However, challenges that tempered these successes are discussed in the next section.

CHALLENGES TO THE HANFORD OPENNESS WORKSHOPS

Institutionalizing Openness

Despite their many accomplishments, the Hanford Openness Workshops faced a variety of challenges, one of which was the difficulty of institutionalizing openness. Changing bureaucratic culture does not happen quickly. Concern for national security and protection of information is deeply ingrained in the institutional structure of DOE. For instance, DOE's different operational activities were separated functionally and financially, resulting in a very compartmentalized, or "stovepiped," administrative structure. Even within the same facility, such as Hanford, different groups may conduct very similar tasks, but may have little or no communication among them (NRC, 1995). Although some gains have been made to break down these artificial walls, the legacy stubbornly persists.

It is much easier to leave information classified and locked away from interested eyes – it costs money and time to declassify, create keywords and index documents, and more to convert them to electronic format, and maintain indices in computer systems and on the Internet. With constant pressure to reduce site budgets, it can be difficult for decision makers to allocate funds for this kind of work. In the short run, it is much easier to avoid involving stakeholders and Tribes in decisions. Again, considerable resources, time, and effort are required to do so, and uncomfortable conflicts can be a daily part of such activities. However, the Workshops' experience echoes the findings of others (e.g., Beierle, 2002; Israel et al., 2001) in suggesting that face-to-face iterative dialogue leads to better decisions – i.e., decisions that are informed by the collective wisdom of a broad spectrum of perspectives and knowledge. Such dialogue also leads to decisions that can be implemented, because those with a stake in the outcome (i.e., those who may challenge such decisions in the courts if they disagree) have been involved in the decision-making process and are more likely to support it. Given that these difficulties pull DOE away from increased openness and public involvement, there is a need for institutionalizing them by defining clear performance metrics, instituting sustained iterative dialogue such as that demonstrated by the Hanford Openness Workshops, and providing sufficient funds through budget allocations.

Expectations

Another challenge was the high expectations of participants. Many assumed that because the recommendations were developed in dialogue with such a broad range of parties, they would be adopted by DOE-Richland. While many recommendations were indeed adopted and implemented, some others were not. For example, some participants were not fully satisfied with DOE responses in the area of contractor performance measures. The Workshops recommended that DOE adopt performance measures for openness, along with financial incentives and penalties, in future contracts. A letter from DOE-headquarters endorsed these concepts (Frei, 1997). However, DOE-Richland took the position that financial incentives need not be written into contracts formally (Wagoner, 1998). This example highlights where additional, directed dialogue to understand the remaining issues for all parties could be used in order to determine if such differences in expectations can be resolved.

Emerging Security Concerns

A third major challenge to openness came about during the second Openness Workshop Series (1999–2001) when concerns about national security and possible espionage at DOE's national laboratories emerged. Workshops participants noted the danger of DOE slipping backward on its openness initiatives by incorrectly equating openness with lax security. Indeed, some in the news media and the U.S. Congress suggested that "too much openness" led to foreign capture of nuclear weapons designs. However, as the Secretary of Energy's Openness Advisory Panel has said, openness is about putting higher fences around a narrower range of information, thereby increasing security. Workshops participants wholly supported the view that DOE strengthens its ability to secure sensitive information to the extent that it earns public trust through accountability and openness. As Joseph S. Mahaley, Director of the Department's Office of Security Affairs, put it:

> Openness provides both a means to assure that the proper sensitive information is protected by classification but at the same time seeks to provide Departmental

accountability to the public by not classifying information that does not have a demonstrable connection to national security (Mahaley, 1999).

Workshops' participants identified three key issues that led to the security problems and leakages in 1999: first, fragmentation of responsibility and lack of accountability in DOE; second, DOE's failure in defining and building "higher walls" around truly sensitive information; and third, breakdowns in management of electronic, as opposed to paper versions of information. However, this is not to say that certain types of information should not be protected. Workshops' participants identified examples of three major kinds of sensitive information that require protection: (1) Designs for nuclear weapons and delivery systems for weapons, including details about achieving fission of plutonium or highly enriched uranium in a particular configuration; (2) Arrangements to safeguard plutonium and highly enriched uranium from diversion, sabotage, and terrorism; and (3) Location, numbers, and other information about how nuclear weapons are deployed (mostly a Department of Defense matter) (HOW, 1999c).

The NRC promotes a similar strategy to narrow the scope of protected information to that which needs to be secure (NRC, 1999). It also described a number of other reasons for balancing openness and security policies at the national laboratories, such as implementing and mediating nuclear arms reduction agreements, recruiting top-notch scientific personnel, and developing policy in the national security sphere:

> The sphere of information of principal concern to national security does not overlap with the areas in which maintenance of openness is essential; hence a proper policy can and should accommodate both tight secrecy in some areas and openness in others (NRC, 1999, p. 7)

The Hanford Openness Workshops participants, the Secretary of Energy's Openness Advisory Panel, and the NRC all conclude that openness activities help avoid security problems by more clearly identifying truly sensitive information and developing appropriate information control systems for that information, while at the same time making non-sensitive information available to Tribes, stakeholders, and regulators.

CONCLUSIONS

The Hanford Openness Workshops (HOW, 1997–2001) evolved out of a climate of distrust about openness – distrust of the Department of Energy as well as distrust among various members of the stakeholder and tribal communities. Despite this conflict, the Workshops provided a forum in which participants could begin to develop the trust necessary to work together toward common goals. In the context of DOE in general, and at Hanford in particular, simply providing such a forum was a "giant step forward" for both openness and public participation. Bringing the parties together using an informal mechanism to discuss technical issues went a long way to build credibility and trust. The Workshops opened and maintained meaningful dialogue about openness for nearly three years. The dialogue allowed participants to understand each other's perspectives about openness and to develop timely, relevant, and specific recommendations about openness. The Workshops allowed DOE-Richland to make tangible progress in its openness initiatives. Such an iterative process of evaluation and refinement, combined with the continued dedication of individuals and organizations will bring DOE closer to its openness goals. The experience of the Workshops also provides further evidence that stakeholders can understand, discuss, and provide input on highly technical matters, providing they have access to appropriate resources. The Workshops illustrate a successful model of participatory process and dialogue that could be emulated at other DOE facilities, or in other contexts.

Nevertheless, encouraging a culture of openness at all levels of DOE remains a critical challenge. Notions of openness and (mis)trust pervade all interactions among DOE, stakeholders, and Tribes. Many DOE administrators have embraced the openness concept, but 50+ years of secret operations and a strict "command-and-control" management style takes time to transform. The Workshops demonstrated a process to cultivate this type of organizational change. A culture of openness requires stable and sufficient funding, free-flowing and transparent information, clear incentives for openness, and a safety conscious and retaliation-free work environment. A one-time commitment is not equal to a sustained, institutionalized culture of openness. Careful attention is needed to ensure that openness remains a central focus throughout the Department of Energy's former nuclear facilities.

The task of long-term management of legacy waste sites also takes place in this culture of secrecy. Thus, the history provided here is important to consider when developing national and site-specific interactions about stewardship. Moreover, the recommendations from the Workshops are substantively important for long-term management because many of the same issues need to be addressed. Finally, the Hanford community is not unique in its mistrust of DOE – and the approach taken with the Openness Workshops could also be used in the stewardship context. An engaged public is an obvious need for effective long-term management.

EPILOGUE

Following the attacks of September 11, 2001 and the subsequent Patriot Act of 2001, the road towards Openness at DOE facilities is barely visible. Almost immediately, nearly all DOE websites were (at least temporarily) shut down, and many never returned, such as the Hanford Geographic Information System. In addition, nearly half of the URLs that were listed on the Internet Resources fact sheet (HOW, 1998a) developed by the Workshops participants are no longer operable, and the Openness Workshops website itself is no longer available. Parts of the Patriot Act effectively reverse tenets of the Freedom of Information Act, which underpins the concept of openness discussed in this paper. The authors are alarmed by these developments.

In 1998, the Secretary of Energy's Openness Advisory Panel emphasized "responsible" openness. The need for responsible openness and security policies is as high now as it ever was. In fact, the tragic events of September 11, 2001 reinforce the need for openness between agencies and stakeholders to ensure broad understanding of true vulnerabilities, and enable effective communication about them during emergencies. Appropriately high levels of security for information that truly needs to be secret is only possible if DOE is open with the vast majority of its information, that which does not require secrecy and should be shared. In the past, accountability was lost behind a veil of secrecy. Based on this experience, some stakeholders and Tribes fear that today's lack of openness will foster carelessness; that DOE will again ignore vulnerabilities that can lead to serious health and environmental consequences. Thus, to travel further down the road towards openness, we need a sustained, iterative dialogue among DOE, stakeholders and Tribes about the trade-offs among secrecy, openness and the related issues discussed in this paper.

NOTES

1. For the purposes of this paper we define the term stakeholder very broadly, to include public interest groups, regulators, DOE employees, and contractors, people

affected by Hanford cleanup, etc. (see Boiko et al., 1996). Native American Tribes, however, are not included in the term stakeholder because they are sovereign entities that have rights to a full government-to-government relationship with the U.S. Government and its agencies.

2. A "curie" is a measure of radiation. As a point of comparison, approximately 25 Ci were released at Pennsylvania's Three Mile Island plant in 1979.

3. CRESP is a university based national research organization created to work with DOE to improve the risk basis for cleanup of complex contaminated environments (CRESP, 2004).

ACKNOWLEDGEMENTS

The authors recognize and appreciate the efforts of all those who participated in Hanford Openness Workshops. Without the immeasurable dedication and enthusiasm of Yvonne Sherman the Workshops would never have materialized. Special thanks also go to all the Openness Participants: Debi Abramson, Mary Lou Blazek, Thomas Carpenter, Tom Cotton, Paul Davis, Greg deBruler, Dirk Dunning, Norma Jean Germond, Andrew Gordon, Deirdre Grace, Roger Heusser, Russell Jim, Judith Jurji, Diane Larson, Angel McCormack, Gerald Pollett, Donna Prisbrey, Rick Stutheit, Jim Trombold, J.R. Wilkinson, and Thomas Woods. Thanks also to Ellen Chu and the anonymous reviewers for comments on earlier drafts of this paper. The Hanford Openness Workshops were conducted with support from the Consortium for Risk Evaluation with Stakeholder participation, operating under agreement with the U.S. Department of Energy (Award #DE-FC01-95EW55084), and with support from the Department of Ecology to the University of Washington (Award #C0000064 and #C0000127). Support for this paper was received from the Institute of Risk Analysis and Risk Communication at the University of Washington, and from the Center for the Study and Improvement of Regulations at Carnegie Mellon University and the University of Washington. However, any opinions, findings, conclusions, or recommendations expressed herein are those of the authors and do not necessarily reflect the views of these organizations.

REFERENCES

Applegate, J. S., & Dycus, S. (1998). Institutional controls or emperor's clothes? Long-term stewardship of the nuclear weapons complex. *Environmental Law Reporter*, 28 ELR 10631.

Arnstein, S. (1969). Eight rungs on the ladder of citizen participation. *American Institute of Planners Journal*, *35*(July), 216–224.
Behrens, J., & Muller, R. (1993). Supply and demand factors in occupational health: Determinants of self-reported compliance with the work place related requirements of the German work safely law. *Occupational Medicine (Oxford)*, *43*(Suppl 1), S47–S49.
Beierle, T. C. (2002). The quality of stakeholder based decisions. *Risk Analysis*, *22*(4), 739–749.
Birkner, L. R., & Birkner, R. K. (1996a). Building a health and safety learning organization. *Occupational Hazards*, *58*(8), 17.
Birkner, L. R., & Birkner, R. K. (1996b). Shifting the EH&S management systems paradigm. *Occupational Hazards*, *58*(10), 68–69.
Birkner, L. R., & Birkner, R. K. (1996c). Team learning: A tool for maximizing safety and health performance. *Occupational Hazards*, *58*(12), 10.
Boiko, P. E., Morrill, R. L., Flynn, J., Faustman, E., vanBelle, G., & Omenn, G. S. (1996). Who holds the stakes? A case study of stakeholder identification at two nuclear weapons production sites. *Risk Analysis*, *16*(2), 237–249.
Bradbury, J. A., Branch, K. M., & Malone, E. L. (2003). *An evaluation of DOE-EM public participation programs*. Pacific Northwest National Laboratory, PNNL-14200.
Brock, J. (1992). *External third-party review of significant employee concerns: The joint cooperative council for Hanford disputes*. Seattle, WA: State of Washington, Department of Ecology.
Buiter, W. H. (1999). Alice in Euroland. *Journal of Common Market Studies*, *37*(2), 181–209.
Clinton, W. J. (1998). Executive Order 13084 of May 14, 1998, Consultation and Coordination with Indian Tribal Governments. *Federal Register*, *63*(96), 27655–27657.
CRESP. (2004). *Consortium for risk evaluation with stakeholder participation*. New Brunswick, NJ. Accessed January 12, 2004, available at http://www.cresp.org
D'Antonio, M. (1993). *Atomic harvest: Hanford and the lethal toll of America's nuclear arsenal*. New York: Crown Publishers, Inc.
d'Errico, P. (1998). *Sovereignty, a brief history in the context of U.S. 'Indian Law'*. Colorado Springs, CO: The Oryx Press. Accessed February 28, 2004, available at http://www.umass.edu/legal/derrico/sovereignty.html
Dininny, S. (2003). US House approves 2004 Hanford budget. *Associated Press State & Local Wire*. Yakima, Washington, DC: November 19. BC Cycle.
Drew C. H. (2002). *The decision mapping system: promoting transparency of long-term environmental decisions at Hanford*. Thesis, Geography. University of Washington, Seattle, WA.
Drew, C. H., Grace, D. A., Silbernagel, S. M., Hemmings, E. S., Smith, A., Griffith, W. C., Takaro, T. K., & Faustman, E. M. (2003). Nuclear waste transportation: Case studies of identifying stakeholder risk information needs. *Environmental Health Perspectives*, *111*(3), 263–272.
Drew, C. H., & Nyerges, T. L. (2004). Transparency of environmental decision making: A case study of soil cleanup inside the Hanford 100 Area. *Journal of Risk Research*, *7*(1), 33–71.
Drew, C. H., Nyerges, T. L., & Leschine, T. M. (2004). Promoting transparency of long term environmental decisions: The Hanford decision mapping system pilot project. *Risk Analysis*, *24*(6), 1641–1664.
Frei, M. W. (1997). *Letter from acting Deputy Assistant Secretary for waste management, U.S. Department of Energy to the Hanford openness workshops*. Washington, DC: U.S. Department of Energy.

Goldberg, S. (1994). *Public access to government information.* Washington, DC: U.S. Information Agency. Freedom Paper No. 6.

Gordon, A. C., Heinz, J. P., Gordon, M. T., & Divorski, S. W. (1973). Public information and public access: A sociological interpretation. *Northwestern University Law Review, 68*(May–June), 280–308.

HAB. (1995). *Hanford Advisory Board advice on the creation of an Openness Panel.* Richland, WA: Hanford Advisory Board. Advice #37. Available at http://www.hanford.gov/boards/hab/

HAB. (1998). *Hanford Advisory Board advice on Hanford openness workshops.* Richland, WA: Hanford Advisory Board. Advice #89. Available at http://www.hanford.gov/boards/hab/

Hanford Education Action League (HEAL). (1992). A sordid sorcery: The history of Hanford's deception. *HEAL Perspective.* Numbers *10/11*(Summer/Fall).

Hanford Reach. (1997). *Single-shell tank waste is in groundwater.* Richland, WA: Hanford Reach.

Hanford Openness Workshops (HOW). (1997a). *Charter for the Hanford Openness Workshops.* Seattle, WA: Hanford Openness Workshops (HOW). Available at http://www.hanford.gov/boards/openness/charter.htm

HONOR. (1998). *Sovereignty (fact sheet for congressional briefing packet).* Bayfield, WI: HONOR-Honor Our Neighbors Origins and Rights.

HOW. (1997b). *Ground rules for the Hanford Openness Workshops.* Seattle, WA: Hanford Openness Workshops (HOW).

HOW. (1997c). *Letter from the Hanford Openness Workshops to Secretary of Energy Frederico Pena regarding performance measures for openness.* Seattle, WA: Hanford Openness Workshops (HOW). Accessed February 2004, available at http://www.hanford.gov/boards/openness/trac-0828/pdf/appendix15.pdf

HOW. (1998a). *Hanford Openness Workshops fact sheet: Internet resources.* Seattle, WA: Hanford Openness Workshops (HOW). HOW-980815–5. Available at http://www.hanford.gov/boards/openness/factsheets/pdf/intrnt_resrcs.pdf

HOW. (1998b). *Hanford Openness Workshops website.* Richland, WA: US Department of Energy–Richland Operations Office. Accessed January 12, 2004, available at http://www.hanford.gov/boards/openness/factsheets.htm

HOW. (1998c). *Hanford Openness Workshops: October 1997–May 1998. Final Report.* Seattle, WA: Hanford Openness Workshops (HOW). TRAC-0818, Rev. 0.

HOW. (1999a). *1999 Hanford Openness Workshops: Summary of Workshop #1.* Richland, WA: Hanford Openness Workshops (HOW). 1.

HOW. (1999b). *1999 Hanford Openness Workshops: Summary of Workshops #2 and #3.* Richland, WA: Hanford Openness Workshops (HOW). 2 and 3.

HOW. (1999c). *Hanford Openness Workshops fact sheet: Openness and security.* Seattle, WA: Hanford Openness Workshops (HOW). HOW-990715-3. Available at http://www.hanford.gov/boards/openness/factsheets/pdf/open_security.pdf

HOW. (1999d). *Hanford Openness Workshops fact sheet: Tribal openness concerns.* Seattle, WA: Hanford Openness Workshops (HOW). HOW-990715-1. Available at http://www.hanford.gov/boards/openness/factsheets/pdf/tribal.pdf

HOW. (1999e). *Is openness working? A progress report.* Seattle, WA: Hanford Openness Workshops (HOW). HOW-991015-1, TRAC-0828, Rev. 0. Available at http://www.hanford.gov/boards/openness/trac-0818/index.html

HOW. (1999f). *Letter from the Hanford openness workshops to Secretary of Energy Bill Richardson and DOE–Richland Field Office Manager Keith Klein regarding the Hanford declassification project*. Seattle, WA: Hanford Openness Workshops (HOW). Accessed February 2004, available at http://www.hanford.gov/boards/openness/trac-0828/pdf/appendix12.pdf

HOW. (1999g). *Letter from the Hanford Openness Workshops to Secretary of Energy Bill Richardson regarding incentives to mediate and zero tolerance for retaliation*. Seattle, WA: Hanford Openness Workshops (HOW). Accessed February 2004, available at http://www.hanford.gov/boards/openness/trac-0828/pdf/appendix9.pdf

Huntoon C. (1999). *Letter from Carolyn Huntoon, Assistant Secretary for Environmental Management, Department of Energy, to Mary Lou Blazek, Hanford openness workshops spokesperson*. Washington, DC: U.S. Department of Energy. Available at http://www.hanford.gov/boards/openness/pdf/1998DOE-HQResponse.pdf

IAP2. (1999). *Foundations of public participation, student workbook*. International Association for Public Participation (IAP2).

IAP2. (2000). *The IAP2 public participation spectrum*: International Association of Public Participation. Accessed February 11, 2002, available at http://www.iap2.org

Israel, B., Lichtenstein, R., Lantz, P., McGranaghan, R., Allen, A., Guzman, J. R., Softley, D., & Maciac, B. (2001). The Detroit community-academic urban research center: Development implementation and evaluation. *Journal of Public Health Management and Practice*, *7*(5), 1–19.

Jim, R. (1994). Nuclear secrecy's legacy: Dislocating native peoples and destroying lands and heritage. Paper presented at comments for seminar one of openness and secrecy: A symposium on establishing accountability in the nuclear age (Russell Jim, Yakama Indian Nation in Washington, DC. May 18–19.

Jorgensen, A. (1991). Creating changes in the corporate culture: case study. *American Association of Occupational Health Nurses Journal*, *39*(7), 319–321.

Kinney, A. G., & Leschine, T. M. (2002). A procedural evaluation of an analytic-deliberative process: The Columbia river comprehensive impact assessment. *Risk Analysis*, *22*(1), 83–100.

Kleinman, D. L. (Ed.) (2000). *Science, technology and democracy*. Albany: State University of New York Press.

Lynch, J. (1999). Raising hell at Hanford. *Seattle Times Magazine*. Seattle, WA. October 17.

Mahaley, J. S. (1999). *Department of Energy memorandum regarding clarification: Department of Energy not ending its openness efforts*. Washington, DC: U.S. Department of Energy, Office of Security Affairs.

McSween, T. E. (1995). *The values based safety process: Improving your safety culture with a behavioral approach*. San Francisco: Van Nostrand Reinhold.

National Research Council (NRC). (1994). *Building consensus through risk assessment and management of the Department of Energy's environmental remediation program*. Washington, DC: National Academy Press.

NRC. (1995). *Improving the environment: An evaluation of DOE's environmental management program*. Washington, DC: National Academy Press.

NRC. (1996). *Understanding risk: Informing decisions in a democratic society*. Washington, DC: National Academy Press.

NRC. (1999). *Balancing scientific openness and national security controls at the nation's nuclear weapons laboratories*. Washington, DC: National Academy Press.

NRC. (2000). *Long-term institutional management of U.S. Department of Energy legacy waste sites*. Washington, DC: National Academy Press.

NRC. (2003). *Long-term stewardship of DOE legacy waste sites-a status report*. Washington, DC: The National Academies Press.

O'Leary, H. (1994). *Guidance on implementation of the department's public participation policy: critical policy elements*. Washington, DC: U.S. Department of Energy Policy (DOE P 1210.1), Public Participation.

Openness Advisory Panel. (1998). *Openness Advisory Panel of the Secretary of Energy Advisory Board: meeting summary*. Washington, DC: Openness Advisory Panel (OAP). (Note that this summary was included as Appendix 6 to the 1998 Openness Workshop Report and is available at http://www.hanford.gov/boards/openness/trac-0818/index.html)

Parker, H. M. (1948). Columbia river situation – a semi-technical review. HW-32809.

Probst, K. N., & Lowe, A. I. (2000). *Cleaning up the nuclear weapons complex: Does anybody care?* Washington, DC: Resources for the Future.

Reason, J. (1995). Understanding adverse events: Human factors. *Quality in Health Care*, *4*, 80–89.

Renn, O., Webler, T., & Wiedemann, P. (1995). The pursuit of fair and competent citizen participation. In: O. Renn, T. Webler, & P. Wiedemann (Eds.), *Fairness and competence in citizen participation: Evaluating models for environmental discourse*, (Vol. 10). Dordrecht: Kluwer Academic Publishers.

Robkin, M. A. (1995). The green run source term study: Special report of the Hanford dose reconstruction project. Washington State Department of Ecology.

SEAB. (1993). *Final report: Earning public trust and confidence: Requisites for managing radioactive wastes*. Washington, DC.: Secretary of Energy Advisory Board (SEAB) Task Force on Radioactive Waste Management.

Stiglitz, J. E. (1999). On liberty, the right to know, and public discourse: The role of transparency in public life. *Oxford Amnesty Lecture*. Oxford, UK, January 27. The World Bank. Available at http://www.worldbank.org/html/extdr/extme/jssp012799.htm

TPA. (2002). *Community relations plan for the Hanford federal facility agreement and consent order*. Richland Washington: Washington State Department of Ecology, U.S. Department of Energy, U.S. Environmental Protection Agency. Available at http://www.hanford.gov/crp/toc.htm

TPA. (2003). *Hanford federal facility agreement and consent order, The "Tri-Party Agreement" (as amended through September 1, 2003)*. Richland, WA: Washington State Department of Ecology; United States Environmental Protection Agency; United States Department of Energy. Available at http://www.hanford.gov/tpa/coverpg.htm

United States Department of Energy (US DOE). (1995a). *Estimating the cold war mortgage: The 1995 baseline environmental management report ("BEMR")*. Washington, DC: U.S. Department of Energy, Office of Environmental Management. DOE/EM-0232.

US DOE. (1995b). *DOE public participation policy*. Washington, DC: U.S. Department of Energy. Policy: DOE P 1210.1. Available at http://web.em.doe.gov/public/empubpar.html

US DOE. (1995c). *Human radiation experiments: The Department of Energy roadmap to the story and the records*. Washington, DC: U.S. Department of Energy, Office of Human Radiation Experiments. Available at http://tis.eh.doe.gov/ohre/roadmap/roadmap/index.html

US DOE. (2000). *DOE OpenNet Database*. Washington, DC: U.S Department of Energy, Office of Scientific and Technical Information. Accessed January 12, 2004, available at http://www.osti.gov/opennet/

US DOE-Richland. (1986). *DRAFT environmental impact statement: Disposal of Hanford defense high-level, transuranic and tank wastes, Hanford Site, Richland, Washington*. Richland, WA: U.S. Department of Energy. DOE/EIS-0113-D.

US DOE-Richland. (1997). *Public involvement desk reference*. Richland, WA: U.S. Department of Energy, Richland Operations Office, Office of External Affairs.

US DOE-Richland. (1998). *Response to the Hanford openness workshops October 1997 – May 1998 Final Report*. Richland, WA: Department of Energy-Richland Operations Office (DOE-Richland).

US DOE-Richland. (2004). *Hanford declassified document retrieval system*. Richland, WA: U.S. Department of Energy, Richland Operations Office. Accessed January 9, 2004, available at http://www2.hanford.gov/ddrs/index.cfm

US EPA. (2003). *Public involvement policy of the U.S. Environmental Protection Agency*. Washington, DC: U.S. Environmental Protection Agency, Office of Policy, Economics and Innovation. EPA 222-B-03-002.

US GAO. (2003). *Department of Energy: reimbursement of contractor litigation costs*. Washington, DC: U.S. General Accounting Office. GAO-04-148R. Available at http://www.gao.gov:8765/cs.html?url = http%3A//www.gao.gov/new.items/d04148r.pdf&qt = GAO-04-148R&col = legal%reports%text&n = 1.

University of Washington. (2004). *Community-based research principles*. Seattle, Washington: University of Washington School of Public Health and Community Medicine. Accessed February 27, 2004. Available at http://sphcm.washington.edu/research/community.asp

Wagoner, J. (1998). *Letter from John Wagoner, Manager of the DOE-Richland Field Office to Mary Lou Blazek, Spokesperson of the Hanford Openness Workshops. Richland*, WA: U.S. Department of Energy, Richland Operations Office. Accessed February 2004. Available at http://www.hanford.gov/boards/openness/trac-0828/pdf/appendix6.pdf

Yin, R. K. (1993). *Applications of case study research*. Thousand Oaks, CA: Sage Publications.

COMPETING PERSPECTIVES ON A PROCESS FOR MAKING REMEDIATION AND STEWARDSHIP DECISIONS AT THE ROCKY FLATS ENVIRONMENTAL TECHNOLOGY SITE

Seth Tuler and Thomas Webler

ABSTRACT

Agencies responsible for remediation and long-term stewardship of areas with chemical and radiological contamination are feeling the pressure to increase public participation in decision-making. Much of the literature outlining advice for how best to involve the public in collaborative decision-making implicitly assumes that there is one best design for such processes.

We report on an empirical investigation into what participants in a process to establish a standard for remediation of plutonium in soil around the Rocky Flats facility near Denver, Colorado think about the most appropriate way to conduct such a decision-making process with public participation. Tapping subjective beliefs and preferences with an approach called Q methodology, we collected in-depth qualitative and quantitative data from 12 experienced participants and agency staff. Analysis of these

Long-Term Management of Contaminated Sites
Research in Social Problems and Public Policy, Volume 13, 49–77

ISSN: 0196-1152/doi:10.1016/S0196-1152(06)13002-1

data revealed three distinct perspectives on what would be the ideal decision-making process for this context. Two of the perspectives emphasized the need to link remediation and stewardship planning, while the third was characterized by the view that these are distinct, sequential activities.

Planners should assume that there may be multiple ideas about what is the most appropriate public participation process for a given situation. Continuing disagreement about the need to link remediation and stewardship can be reflected in disputes about process design. Success should be viewed as a function not only of the design features used but also the extent to which the design matches the needs and preferences of the participants.

INTRODUCTION

Chemical and radiological contamination at many Unites States' government facilities and former industrial sites is both very extensive and long-lived. This situation is particularly extreme at those sites that have played a role in the production, testing, and research of nuclear weapons. Efforts to contain and manage hazards at these sites use a three-step sequential approach entailing contaminant reduction, isolation of remaining contaminants, and long-term stewardship activities (National Research Council, 2000). Because of the difficulty of removing hazards completely, timely and effective planning for stewardship based on long-term institutional management has become a pressing national problem (National Research Council, 2000, 2003; Department of Energy, 1999, 2001a).

For the most part, the Department of Energy (DOE) has behaved as if "cleanup" (or contaminant reduction and isolation) are activities independent of stewardship and, furthermore, that stewardship is something that happens after these remediation activities (National Research Council, 2003). However, remediation actions are closely intertwined with decisions about long-term stewardship (National Research Council, 2000, 2003). Decisions about how much contamination can be left in "cleaned" soils or groundwater have implications for future plans to manage risks to public health from remaining contamination (Applegate & Sarno, 1997; Department of Energy, 1999; Fernald Citizens Task Force, 1995; Rocky Flats Stewardship Working Group, 2001). Examples of such connections were repeatedly encountered in decisions about criteria for waste acceptance at Washington State's Hanford site to on-site disposal at the Fernald facility in Ohio to standards for acceptable residual plutonium contamination in soils around the Rocky Flats facility in Colorado; at the latter two sites "cleanups" were completed in 2006.

Furthermore, in its "Phase 2" report the Committee on Long-Term Institutional Management of DOE Legacy Wastes (National Research Council, 2003) wrote that "how planning and cleanup are carried out shapes the confidence the public, stakeholders, and political leadership will place in DOE as cleanup ends. Not only is the substance of LTS affected by choices made in the cleanup process, but so is the social setting in which LTS will be conducted. That setting is critically important to the ability of the steward to discharge its responsibilities" (p. 27). In short, the report argues that *processes* of planning and decision-making for remediation will be important to the success of long-term institutional management. Thus, the Committee also discussed the importance of community involvement for promoting trust between a community (including local governments) and the DOE, for crafting decisions that are accepted as legitimate, and for identifying options that build on the capacities of the local community. And, it provided suggestions related to promoting good community involvement, such as the need for "early, continuous involvement of stakeholders' advisory groups with frequent contact, complete candor, rapid, and full response" (p. 28).

In fact, a growing body of research has identified features for public participation that can be important for achieving meaningful, legitimate, and timely outcomes (Ashford & Rest, 1999; Beierle & Cayford, 2002; Bradbury, Branch, & Malone, 2003; Chess & Purcell, 1999; Drew et al., 2003; National Research Council, 1996; Webler & Tuler, 2000). Many of these are now reflected in policy guidance from federal and state agencies (e.g., Department of Energy, 2003; Environmental Protection Agency, 2003) and proposals for best practices from practitioners, nongovernmental organizations, and others (Environmental Law Institute, 2001; Environmental Protection Agency, 1996; Ledwidge, Moore, & Crawford, 2004; Till & Meyer, 2001). Unresolved, however, is the question of how to weave together such advice to design a process for a *particular* context.

In this paper, we explore this question in the context of efforts to establish standards for removal of radionuclides from soils around the Rocky Flats facility near Denver, Colorado. This standard affects the levels of residual risk to future visitors and users of the site because it determined the radiological contamination that was allowed to remain in the soils after remediation was completed and during long-term stewardship (Risk Assessment Corporation, 2000; Institute for Energy and Environmental Research, 2001). We asked people who participated in the process to set the soil standards about their notions of the most appropriate way to conduct such a decision-making process. We found that people expressed different preferences and they placed different emphases on the need to integrate remediation and stewardship planning.

Our findings are relevant to the larger mission of promoting public long-term involvement in risk management and long-term stewardship. First, planners should assume that there can be multiple ideas about what is the most appropriate process for a given situation. Second, the findings highlight the ways that continuing disagreement about the need to link remediation and stewardship can be reflected in disputes about process design. Processes for assessment and decision-making during remediation that do not meet the needs of stakeholders may undermine the quality and legitimacy of hazard management decisions and the capacity of organizations to meet their responsibilities for long-term stewardship.

BACKGROUND

In our research, we ask people what they think are important features of a decision-making process based on their prior experiences and involvement with an issue. Thus, before discussing our research methods and findings we provide a brief overview of the context underlying our research and informing our subjects' experiences. The Rocky Flats Environmental Technology Site is located about 15 miles northwest of Denver, Colorado. Rocky Flats is the location of a former nuclear weapons production facility owned and operated by the DOE. It began operations in 1952, with the primary mission of producing nuclear weapons' components and assemblies (i.e., plutonium pits) manufactured from uranium, plutonium, beryllium, and other metals (Department of Energy, 2001b). Both routine operations and accidents resulted in widespread contamination of the site with radionuclides and chemical substances. In 1989, the site was placed on the National Priorities List established as part of the Comprehensive Environmental Response, Compensation, and Liability Act (CERCLA, also known as Superfund).

From 1994 to 2006 the mission of Rocky Flats was site contaminant remediation and isolation. Kaiser-Hill was the contractor responsible for these activities. DOE and Kaiser-Hill finished remediation-related activities during 2006, at which time long-term stewardship activities began. As part of this plan, most of the site became a National Wildlife Refuge managed by the United States Fish and Wildlife Service; about 1,000 acres of the most contaminated parts of the 6,500-acre site will be managed by the DOE's Office of Legacy Management. Many stakeholders continue to question the adequacy of the remediation effort and the long-term stewardship plans to protect human health.

Plans for long-term institutional management of the site are critically dependent on decisions that were made about remediation, including the levels of contaminants allowed to remain onsite. Thus, during the last several years of the remediation effort DOE, its regulators, the contractor, and the community focused on the levels of residual plutonium that will remain in soils after contaminant remediation and isolation are completed (Institute for Energy and Environmental Research, 2001; Earle, 2004; Risk Assessment Corporation, 2000; Rood, Grogan, & Till, 2002; Till & Meyer, 2001). In 1996, DOE and its regulators, the Environmental Protection Agency and Colorado Department of Public Health and Environment, integrated interim "radionuclide soil action levels" (RSALs) into the Rocky Flats Clean-up Agreement for (a) plutonium (239 and 240) and its associated americium-241 and (b) uranium isotopes (Department of Energy, US Environmental Protection Agency, and Colorado Department of Public Health and the Environment, 1996). The Rocky Flats Clean-up Agreement is the legal agreement between DOE and its regulators that governed contaminant remediation and isolation activities at the site. RSALs are a legal measure used to determine whether radioactive materials in soil must be removed or contained to reduce risk. If contamination was below the RSAL no remedial action need be taken.

After the interim RSALs were determined, various stakeholder groups expressed concern that they would not provide adequate protection of public health. Most of the attention was placed on the RSAL for plutonium, which was proposed to be 650 pCi/g of soil. The Rocky Flats Citizens Advisory Board (RFCAB),[1] Rocky Flats Peace and Justice Center, Town of Westminster, City of Broomfield,[2] local Congressman David Skaggs, and others asked for an independent assessment of the proposed RSALs and how they were determined before final values were selected.

In response, the DOE funded an independent community-based assessment of the interim RSALs in 1998. This assessment was conducted by a committee called the Rocky Flats Radionuclide Soil Action Levels Oversight Panel (RSALOP). The Panel comprised 13 individuals representing various technical, community, and local government perspectives. In addition, it provided opportunities for involvement and outreach with the broader community concerned with Rocky Flats cleanup, closure, and future use. For example, monthly working meetings were open to the public and a series of public workshops were held (RFCAB, 2001). RSALOP hired an independent contractor, Risk Assessment Corporation (RAC), to assess the interim standards. Much attention was given to questions about the validity of assumptions and assessments used to determine the interim RSALs and about their effectiveness at protecting public health over long periods of time. In fact, the interim RSAL for plutonium at Rocky Flats was higher than levels set for other sites (RAC, 1999). RAC completed its work for the RSALOP in 2000 and recommended lowering the RSAL for plutonium (and its associated americium-241) from the interim value of 650 pCi/g to 35 pCi/g of soil to meet public health protection goals (RAC, 2000).

Many have felt that the RSALOP process was a success (Earle, 2004; Till & Meyer, 2001) although there continued to be controversy about its recommendations. For example, a technical review of RAC's assessment was conducted by the Institute for Energy and Environmental Research (2001) on behalf of the Rocky Mountain Peace and Justice Center, with a particular focus on whether proper assumptions about potential exposures from future land use were used. They concluded that the RSAL for plutonium should be reduced to no more than 10 pCi/g of soil because of a variety of factors, including the risk of fire on the site and speed of potential plutonium migration into groundwater. Ultimately, the RSALOP recommended the RAC values to the DOE and the regulators and the DOE adopted the recommended values.

While the RSALOP process was ongoing, two other groups played key roles in the establishment of the RSALs. The community advisory board (RFCAB) co-sponsored workshops and provided opportunities for public education. The Rocky Flats Coalition of Local Governments (RFCLOG)[3] provided comments and recommendations about computer modeling assumptions and parameters, while working on future use and long-term protection issues, worker protection and health issues, and facilitating communication between state and federal agencies and local elected officials.

At the same time RSALs were being proposed, evaluated, criticized, and discussed, other important issues related to cleanup, closure, and future use were also being considered. These included

- proposals to transfer ownership and responsibility for much of the site to the United States Fish and Wildlife Service,
- determinations about possible future uses of the site,
- decisions about waste transportation, and
- public health studies about historical releases of contaminants offsite.

In particular, the RFCAB and RFCLOG provided comments to the DOE expressing the importance of considering the RSALs as just one component of a larger effort to ensure that Rocky Flats would be satisfactorily remediated and safe, long-term stewardship be ensured. For example, the RFCAB made a recommendation that the RSALs be revisited at a later time when new science and remediation technologies become available. The RFCLOG noted that RSALs, as proposed, will not ensure that other water-quality regulations are met.

This brief history sketches out the landscape of the decision-making process upon which our research study focused. When we inquired of people about their preferences regarding process, they responded in the context of this mosaic of activities and players.

RESEARCH METHOD AND QUESTION

We sought to inquire if people who participated in the process to establish soil action levels had different expectations for what the process should be. Toward this end, we employed Q methodology (Brown, 1986, 1996; Focht, 1995; McKeown & Thomas, 1988). "Q" is a type of discourse analysis that uses quantitative and qualitative methods to reveal in-depth cultural perspectives on a subject. It taps the subjectivity of people with different experiences, interests, and values to reveal their opinions; in this case, we used it to understand their opinions about what would be the ideal collaborative decision-making process for establishing RSALs for Rocky Flats.

Q methodology has subjects react to statements related to an issue of interest. In this study, each statement represents a building block for a perspective on what is an ideal decision-making process. Researchers assemble a set of statements that they presume will supply all the ingredients necessary for the subjects to express their personal beliefs. A cooking metaphor is helpful. The researchers supply a kitchen full of ingredients and then ask each individual to cook their favorite dish using all the ingredients. Everyone will produce a unique outcome, even though they are required to use all the ingredients, because they can select the amount used and the way they are combined.

Selecting the Q statements is of obvious importance. Since we cannot presume to know all the existing perspectives (what dish each person will make), we selected statements to be as widely inclusive as possible. Statements came from (1) interviews we conducted with participants in other studies (we did not interview people in this case) and (2) a thorough review of the academic and practitioner publications in the field. From these two sources, we extracted ideas and claims about public participation in environmental and risk-assessment decision-making. We coded statements into conceptual categories and sampled from each category to ensure we had a set of statements that was broadly representative.[4] The full list of "Q statements" is given in Table 1. Typically, researchers select no more than four or five dozen statements because getting people to sort more than this number is difficult.

To conduct the Q sort exercise, we handed each subject a set of small cards (about the size of a normal business card), each with one Q statement printed upon it. A "condition of instruction" specified the context under which the participant was to interpret and react to the Q statements. It was

> Imagine that the DOE's process to set clean-up levels for contaminated soils at the Rocky Flats site is going to be done again. Sort the statements according to what you believe should be the most important to least important factors guiding the design of the process.

This condition of instruction was designed to focus the participant's thinking on the topic of remediation and isolation of radiological contamination in soil around Rocky Flats and the setting of RSALs specifically. We wanted to draw on the participant's experience with the decision-making and public participation processes to date and at the same time get his or her ideas of what would be the best way to design such a process in the future. We did not ask people to evaluate the process that had occurred, although we expected, of course, that their experiences would inform their ideas for a new process.

This is how the Q sorts happened. We asked each subject to read all the statements through once or twice. Then we asked them to sort the statements into three piles, the left-hand pile being the less important ideas, the right-most pile being the most important ideas, and the middle pile being in between. We then asked them to continue sorting the statements according to their relative importance in their individual opinion. The sorting was constrained by forcing participants to sort the cards into a specific pattern that forms a normal distribution.[5] There were 11 categories for sorting and the ends were anchored with *most important to me* to *least important to me*. To realize a normal distribution of the data, the number of cards allowed in each of the 11 columns was 3, 3, 4, 6, 7, 10, 7, 6, 4, 3, and 3. A subject may have felt that *all* the statements were important, but he or she still had to differentiate between the *most* and *least* important. Notice that, while the right-most edge contains statements the participant thought were most important, and the left-most edge contained statements considered least important, the middle *does not* contain statements that are viewed as irrelevant or unimportant.

Table 1. Q Statements and Their Rankings in Each Factor Emerging from the Analysis. Factors A, B, and C are Characterized by a Particular Ranking of the Q Statements into 11 Categories from "Least Important" (–5) to "Most Important" (+5).

No.	Statement	Factors Emerging from Analysis		
		A	B	C
1	Set up a situation that encourages all participants to listen to what others say and to consider it carefully	1	4	−1
2	Use the best available science in the analysis	5	1	5
3	Establish relationships that promote constructive collaboration among participants	−2	0	4
4	Acknowledge and explore uncertainties	1	1	1
5	Develop a common language and understanding among participants	0	1	−2
6	Reach out in a number of different ways through different mechanisms to different communities on different issue points, throughout the process	0	−1	−1
7	Work to build trust among the different participants during the process	−1	0	3
8	Hold meetings at different times and places so no one is excluded from participating	−3	3	−3
9	Participants should be courteous and respectful to one another	2	0	0
10	Provide financial resources that enable people to participate effectively (e.g., travel, hire experts)	0	−1	−2
11	Participants should see beyond their individual interests to what is good for the larger community	3	1	3
12	The process cannot be open to just anyone who wants to participate, participation has to be restricted in some way	0	−5	−5
13	Participants should be accountable for what they say, sincere in their promises, and reliable in carrying them out	−2	0	4
14	The process gives recommendations to the DOE, who then makes the final decisions	−1	−5	−1
15	Participants should have reasonable expectations about what the agencies are able to do	1	−4	4
16	All important decisions are made according to consensus (including the agenda)	−5	5	−4
17	Participants should attend meetings regularly and see tasks through to completion	3	0	0
18	It is clear under what conditions the process will end	2	−2	−2
19	Participants should be able to deal with complex technical issues	1	0	2

Table 1. (*Continued*)

No.	Statement	Factors Emerging from Analysis		
		A	B	C
20	Every recommendation is justified with evidence	−1	−3	3
21	Participants should feel comfortable and safe at the meetings	−2	1	−3
22	Consensus is used to decide what rule is used to make decisions (simple majority vote, 2/3 majority vote, etc.)	−4	3	−4
23	There are clear ground rules that govern how people should interact	4	3	1
24	The DOE responds in a timely way to all questions, comments, and requests	2	2	2
25	Pay attention to the physical arrangement of tables and chairs at the meetings	−5	−1	−5
26	Opportunity cannot be an empty shell; there need not only be opportunities to be heard but there also has to be some way for the public to see that the decision-makers are listening	1	2	0
27	Discuss the values underlying people's opinions about the issues	−2	−1	−3
28	There are mechanisms for communicating to the broader public about what decisions are being considered and made	−1	0	0
29	Validate all information to make certain it is correct	−1	−2	2
30	Participants who represent groups check in with their memberships regularly to ensure that they represent their views accurately	−4	−1	−3
31	Everyone has an equal chance to put their concerns on the agenda	−3	5	0
32	The process improves the participants' skills to participate effectively in processes like this (e.g., problems solving, conflict resolution, communication)	−2	−2	−5
33	The process has to be able to limit topics of discussion in order to avoid quagmires	0	−2	0
34	The process improves participants' understandings	−1	−2	0
35	The process requires unbiased and independent facilitation	4	4	0
36	The process ends up enhancing the trust between the community and DOE	−1	−2	5
37	The purposes and goals of the process are clear to all involved	4	2	2
38	The process does not make any pre-existing conflicts worse	−3	−3	−2
39	All participants have equal access to information	2	2	0

Table 1. (*Continued*)

No.	Statement	Factors Emerging from Analysis		
		A	B	C
40	All important stakeholders are taking part in the process	5	3	−1
41	There is full disclosure of information at all times	3	5	1
42	At the end of the process there is a clear plan for how to implement the final decision	5	−1	0
43	The staff involved are receptive to questions or requests for information from the public	0	0	1
44	The process makes progress on solving the right problem	3	0	3
45	Get the right information	0	0	5
46	The process produces outcomes that are acceptable to me or my organization	−4	−4	−1
47	The process taps the knowledge and experiences of local people	2	1	−1
48	The process produces outcomes that are acceptable to the DOE	−5	−5	−4
49	The process needs an effective leader	1	−4	2
50	One outcome of the process is a plan to ensure that the promises made are actually followed through, that organizations are accountable for their promises	1	−3	2
51	There is adequate administrative support (e.g., funding, staffing) for the life of the process	2	1	−2
52	The process is well timed to the DOE's window of opportunity to act	0	−3	1
53	There is adequate notification of meetings, comment periods, etc.	0	4	−2
54	Allow time to re-visit issues and decisions, even if it means extending the timetable	−2	−1	−1
55	Participants are involved in deciding *what* studies ought to be done	0	2	1
56	Participants are involved in deciding *how* studies ought to be done	−3	2	1

We selected 12 people who participated in the RSALOP, RFCAB or RFCLOG, or who worked for DOE or Kaiser-Hill to complete Q sorts. In addition, we endeavored to select individuals who would represent unique perspectives regarding public participation in decision-making. One individual worked for the DOE and another for Kaiser-Hill. Three were local government officials or staff from nearby communities. Two were members of advocacy groups concerned with cleanup, environmental health, and disarmament issues arising from Rocky Flats (as well as more broadly). Four considered themselves to be community members, but did not affiliate themselves with any particular advocacy group. One was a staff person for an advisory group. All of them had been long-time participants in processes related to Rocky Flats.

An advantage of Q method is that only a small number of research participants are needed, if they are chosen carefully. This is because participants do not represent a population in a conventional survey sense. In a Q study, the sample is *not* the people who sort the statements. Rather, the sample is the set of Q statements and the population is the universe of utterances that have been made on the topic. The Q sorts completed by people are the variables. This is just the opposite of standard survey techniques. By tapping the subjectivities of a small group of individuals with unique perspectives (i.e., the variables), it is possible to reveal cultural perspectives. Continuing with our metaphor, if we had 12 people bake their favorite dish using some 50 ingredients we supplied, we would get a fairly good cross-section of all dishes represented in the culture. The population would be all the ingredients used in cooking, the sample would be the 50 ingredients we supplied, and the variables would be the combinations of ingredients into finished dishes. We would then be able to analyze the patterns of ingredient usage to say something about cultural categories of dishes (e.g., desserts, soups/stews, entrées, appetizers, etc.).

We approached each individual via telephone or e-mail, introduced them to the project, described how they were selected, and told them about Q methodology and the Q sort exercise. We visited them at their convenience. The entire process with each person took about one and one-half hours. Data were gathered during June and July 2002. Participants reported the Q sort was innovative, fun, and that it stimulated their thinking.

For data analysis we used a freeware program called MQMethod, which performs a factor analysis upon a correlation matrix. We used factor analysis to reveal patterns in how individual statements were correlated across the different Q sorts and identified each factor with a particular perspective on making decisions about site remediation and stewardship. In this instance, the analysis reduced the 12 sorts (variables) to three factors. Each factor is expressed as an idealized Q sort in the program's output. Thus, each factor is characterized by a particular rank ordering of the Q statements into the 11 categories from "least important" (–5) to "most important" (+5).

In addition, the program computes how closely each individual's sort is related to each factor. For each Q sort a "factor loading score" is computed, in essence telling the degree to which that individual subscribed to a factor. The score ranges from +1.00 (indicating that participant's sort exactly matched the factor) to –1.00 (indicating that participant's sort was the exact opposite of the factor). Zero indicates no similarity at all. In addition to the statistical analysis, we ran an audiotape during the Q sort exercise and recorded the conversation we had with the participant. We asked each participant to interpret their sort and to explain how he or she interpreted specific Q statements. We had these tapes transcribed and consulted them when interpreting the factors.

We started every analysis using principle components analysis followed by the varimax solution. Theoretically this solution accounts for the most variance in the data. It is customary within the Q community of researchers to use this approach. However, some argue that the varimax solution is not necessarily theoretically relevant and that judgmental hand rotation is sometimes needed to find the most appropriate solution. We used judgmental hand rotation after computing a varimax solution because we were able to separate individuals more clearly with respect to their factor loading scores and we were able to reduce the number of people who did not load significantly on any factor. We also deemed the solutions to be meaningful and relevant. We selected our factors based on three criteria.

1. The solution should account for over 50% of the total variance in the data.
2. Each factor solution had to account for at least 10% of the total variance.
3. The factor had to be meaningful and theoretically important.

Each of the three factors represents an idealized social perspective about what is an appropriate process in the context defined by the condition of instruction. The tricky part of the analysis is figuring out exactly what each factor means. Based on the arrangement of statements in each of the three factors, and with the help of the transcribed conversations, we composed a written narrative describing the particular perspective represented by each factor. To validate the narrative's accuracy, we selected the one or two people whose individual factor loading scores were highest for each factor and e-mailed the associated narrative description to them for confirmation. In all cases our narratives were confirmed with minor editorial amendments.

RESULTS

Three distinct and coherent factors emerged from the analysis. Each factor represents a perspective on what would be the most appropriate decision-making process for setting soil action levels to guide remediation and stewardship planning at Rocky Flats. Table 1 lists the statements along with the rankings each statement received in each of the three factors. We labeled the factors, and their associated perspectives, simply by A, B, and C.

Table 2 presents the factor loading scores of each participant and gives the total variance explained by each of the three factors. In the table, we listed subjects according to the factor on which they load most highly. Subjects 1–5 load significantly on factor A, subjects 6–8 load significantly on factor B, and subjects 9–12 load significantly on factor C.

What is particularly important is that every person loads significantly on at least one factor. Only subject 7 loads significantly on two factors. This result suggests that this individual either expressed a point of view that is unique and not captured by any of the three "ideal types," or that this person has genuinely mixed ideas about what is the most appropriate decision-making process with public participation. We investigated other factor solutions through additional judgmental hand rotation of factors and

Table 2. Loading Scores of Each Participant on the Factors A, B, and C[a].

	Individual Loading Scores on Factor		
	A	B	C
Factor A			
Subject 1	**0.76**	0.07	0.13
Subject 2	**0.71**	−0.11	0.14
Subject 3	**0.66**	0.05	0.05
Subject 4	**0.65**	0.17	0.11
Subject 5	**0.65**	0.19	0.02
Factor B			
Subject 6	−0.03	**0.79**	−0.32
Subject 7	0.01	**0.70**	**0.42**
Subject 8	0.41	**0.68**	−0.18
Factor C			
Subject 9	0.06	0.37	**0.73**
Subject 10	0.38	−0.27	**0.72**
Subject 11	0.33	−0.04	**0.68**
Subject 12	0.38	0.22	**0.48**
Variance explained (%)	24	16	18

[a]Significant loading scores are highlighted in bold type. A loading score greater than 0.4165 is statistically significant at the 0.05 level.

inclusion of additional factors, but we discovered all the new solutions had more participants confounded on more than one factor, higher inter-factor correlations, and/or less variance explained. Thus, these alternative solutions were not as informative about the differences in preferences among the participants in our study.

In the following sections, we present the three perspectives that are represented by factors A, B, and C via narratives that describe what kind of process the individuals participating in our research would create if they were responsible for designing a new effort in this case. Since the narratives are constructed from the Q statements, references to important Q statements and their relative rankings in the factors from Table 1 are included in the descriptions. In this discussion, we limit our references to the type of people who ranked highest on each factor, mainly because we are committed to respecting the anonymity of our research participants.

Perspective A

This perspective advocates a responsible cooperative process among key interest groups based in sound science and sensitive to producing feasible outcomes with a reasonable amount of time and effort.

A responsible collaborative process, according to this viewpoint, contains numerous qualities. Foremost is the inclusion of all the important stakeholders (40). For collaboration to be effective and satisfying, there need to be facilitators (35) and clear rules (23). That said, the participants in this vision also have responsibilities. They should agree on the goals of the process (37), attend meetings with regularity (17), see beyond narrow self-interests (11), and be respectful and courteous to each other (9). A responsible process also has pragmatic value in that it results in a clear plan for how to implement the outcomes (42).

While the process should result in well-crafted and relevant outcomes, those holding this viewpoint do not contend that the recommendations have to be acceptable to the DOE (48) or to one's own organization (46). Recommendations do not have to be acceptable to the DOE because there are other important agencies/institutions with responsibilities for managing Rocky Flats during remediation and stewardship. For example, local government will have important responsibilities for ensuring that plans for long-term stewardship are achieved. Also, because this is a process that focuses on the participation of important stakeholders, there is no need to make the process broadly accessible. For example, meetings do not need to be held in different locations or at different times (8). Allowing restrictions on who can participate (12) was weakly emphasized, but ranked much higher on this perspective than either of the others. The lack of attention to broad accessibility is, in part, a result of the process having already gone on for a number of years. There is a feeling that the key stakeholders already understand the concerns of the general public and the nature of others' values.

A second key theme in the perspective represented by factor A is that the process should use the best-available science (2). This aspect ranks a distant second to the insistence on a responsible process. Moreover, there is no need to make all the information widely available because participants should focus on what is important to solving the problem at hand. Thus, it is only of moderate importance that the process taps the knowledge and experience of local people (47), there be full disclosure of information at all times (41), and participants have equal access to information (39).

The third major element of this perspective is a concern with actually reaching "closure" by completing remediation activities and beginning long-term stewardship activities. This is reflected in the very low scores given to statements about relying on consensus (16, 22), on the importance of producing implementable outcomes (42), and the relatively high score given to the statement about needing to be clear about when the process will end (18). There is a concern that people may use the process to stall or prevent action, thus the low level of support for having everyone able to put their concerns on the agenda (31) or to allow time to revisit decisions that have been made (54). Interestingly, there was not support for establishing relationships that promote constructive collaboration among the participants (3). This is consistent with the lack of support for consensus. According to this perspective, it probably will not be possible to have everyone agree. Despite this, decisions need to be made and the process needs to progress.

Perspective B

This perspective describes a process that focuses on maximizing the opportunity of people to participate effectively while also minimizing power imbalances among them. It places relatively little emphasis on science and information. In this viewpoint, participants should be able to influence the outcomes of the process. Consequently, those supporting this perspective are sensitive to issues of access, accountability, and power.

Meaningful access to the process is ensured when everyone is on the same footing. All information needs to be disclosed (41) and participants should have equal access to it (39). There should be adequate notification of meetings (53) and meetings should be held at different times and locations (8). Meetings held only during working hours or in far-away locations (e.g., Denver) are not appropriate because this does not aptly meet the needs of volunteer participants. It is strongly emphasized that people should not be excluded from participating (12). One element of empowering people's participation is if they feel comfortable and safe while participating (21).

Participants are placed on a more equal footing when an independent facilitator (35) is present and there are clear ground rules (23). This general interest in minimizing differences of power and influence among participants is further illustrated by support for ensuring participants have an equal chance to put concerns on the agenda (31). Complementing this is the low importance assigned to limiting topics of discussion to avoid quagmires (33), suggesting that topics should not be limited because it would unfairly constrict people's ability to pursue agenda items they believe important. Thus, for example, those associated with this perspective are critical of DOE efforts to separate remediation and stewardship planning.

Similarly, participants should not be constrained by having "reasonable" expectations about what DOE can do (15), presumably because that could marginalize some views. In fact, there is a sense that the more DOE and Kaiser-Hill are pushed, the more they will do. To further protect the interests and needs of all stakeholders, consensus is very important to this perspective. Consensus should be used to make all important decisions (16), including recommendations, and to determine decision rules (22).

The character of meetings is important to the perspective represented by factor B, because the quality of interactions can affect accessibility and power. There should be ground rules to govern interaction (23). Here the importance is on ensuring that participants feel listened to (1) and that they are able to explore all concerns and make progress on the problems of interest to them. As one high loader on this factor noted, the problems identified by the participants may be different than the problems as defined by government agencies (44).

A third aspect that distinguished perspective B has to do with the role of science and technical information. Both of the other factors emphasized this issue, while this perspective views the process as basically driven by democratic ideals, not giving a privileged role for science and technical information. As evidence for these claims, we note the low scores for using the best-available science (2), for getting the right information (45), for validating information (29), and for making sure that recommendations are supported by evidence (20). According to our interviews with those who loaded highly on factor B, these statements were understood as referring to *technical* information. This is not to say that science and technical information are *not* important – they clearly are. It is just that scientific understandings are always limited.

At the same time, there is a concern that people not be excluded because of lack of technical knowledge. Discussions, reports, and the like should be conducted at a level understandable to the participants. There should be access to independent technical resources to help explain complicated issues and documents and point out areas of concern. As one person who loaded highly on factor B noted in the interview, what is critical is that both technical and nontechnical information be clear, honest, and open about biases. A good process will help participants develop better understandings of technical issues.

Perspective B does not place much emphasis on whether the process produces a plan to ensure that promises are carried out (50) or that the process is well timed to the agencies' opportunity to act (52). There is skepticism about requirements to meet DOE schedules at the expense of having a meaningful dialogue – especially because there is a sense that DOE often fails to meet its own schedules. Furthermore, producing recommendations for the DOE, who then makes the final decisions (14), is viewed as very unimportant. From our interviews with the research participants loading significantly on this factor, we learned they believe that DOE measures success based on the *quantity* of recommendations emerging from the process rather than the *quality* of recommendations. In contrast, they believe that recommendations are important to creating a paper trail that can serve to undermine claims that the public is behind all of DOE's decisions. At the same time, participants want to make progress; it is just that their view of progress may differ significantly from the DOE's or contractor's views. With a view toward the broad, long-term needs of the affected community and creating effective stewardship, there is little desire to meet the more narrow bureaucratic needs of the DOE's or its contractors' focus on remediation.

Perspective C

This perspective depicts a process that is foremost based on science and information. It also does not focus so much on who participates as on the disposition of the participants and the effect that their participation has on the process. It asks that they be cooperative and trusting. A third general feature is a focus on pushing decisions forward, avoiding delays, and working to ensure actions are taken on remediation, without being concerned at this stage on stewardship planning. This is a process that retains the authority of the responsible authorities and allows them to make their own choices, as informed by participants in the process.

The most important aspect of the perspective represented by this factor is that the process should focus on scientific and public health aspects when setting soil action levels for the site. This requires the best-available science (2), getting the right information (45), validating its correctness (29), and justifying recommendations with evidence (20). What is interesting is that this perspective sees the process as fairly closed. For instance, there was not strong support for the statements about ensuring full disclosure of information (41) or for giving participants equal access to information (39). Because the issues are highly technical, the process should avoid simplifying information for those without a firm understanding of the science. Participants should be able to deal with complex technical issues (19).

The second main element of the perspective represented by perspective C addresses the attitudes and behavior of the participants. Enhancing trust between the community and the DOE (and its contractors) (36) is an extremely important goal, as is building trust (7) and constructive collaborative relationships among participants (3). This is especially important when there are conflicting technical experts, uncertain scientific understandings, constantly improving knowledge, and no uniform national standards for soil remediation levels. These goals are also important because the process must result in time-sensitive decisions (44). On this theme, participants made mention that initial decisions must be taken expediently so that the site can be fully "cleaned" by DOE's "accelerated cleanup" deadline of 2006.

What is best for this kind of process is that the participants adopt the right attitude. As part of a credible, science-based, decision-making process, the responsible agency should know what it expects from participants and make those expectations clear to the participants. During our interviews, those associated with this perspective reported they felt that past efforts at Rocky Flats failed to be clear about what was needed from the participants. Thus, they hope that participants will have reasonable expectations about what the agencies can do (15), that they will be accountable for what they say and do (13), and that they will be able to see beyond their self-interests to the larger good (11). There is slight support for an effective leader to keep the process focused (49). There is a feeling that if the DOE, Kaiser-Hill, and the participants can focus on the important problems at hand and stay away from self-interest, then the participants will be able to make reasonable, competent recommendations that promote the broader good.

The third central element of this factor is to ensure the process moves forward. The two key aspects of this are keeping participation fairly restricted and avoiding the use of consensus. Those holding this perspective do not place a strong emphasis on making the process widely accessible. Meetings do not have to be held at different times and locations (8) to facilitate participation, in part because past experience reveals that "core" participants will come regardless of how convenient meetings are and that "John Q. Public" will rarely come no matter how meetings are arranged. The need to ensure that all the important stakeholders take part in the process (40) receives weak support here. The perspective envisions a process in which a small group of dedicated and experienced people take part. This explains the low importance to things like making sure participants feel comfortable and safe (21), providing financial resources to enable participation (10) (their groups have the needed resources), or developing a common language among the participants (5) (because they already have one). Importantly, consensus is viewed as a way of allowing the process to be side-tracked; therefore, it is strongly not supported (16, 22). One person loading significantly on factor C noted that outcomes based on consensus are often "watered down" and have limited value, because they must please so many competing interests.

Comparison of Perspectives Represented by the Three Factors

One of the interesting aspects of Q method is that it enables us to compare and contrast perspectives on the same elements. We consider statements that achieved scores across all three factors within the narrow range of two points on the 11-point scale to be *consensus* statements and those with a wider range of scores as *distinguishing* statements. However, it is also important in Q method not to take too reductionist an approach to the analysis. The story to be told by each factor emerges from the patterns of all the statements.

Consensus Statements

In this project, the consensus statements were not very telling. Nearly all of the consensus statements in this case appeared in the middle of the factors. For example, acknowledge and explore uncertainties (4) was ranked +1 in all factors. The strongest positive consensus statement was that the goals and purposes of the process be clear to all (37). Heeding our own warning above (about considering statements out of context of all the others), we note that each perspective expresses a different idea of what the goals and purposes are. However, it is worth noting that all perspectives express that it is beneficial to have clarity on the goals.

The strongest negative consensus statement was that the process should produce outcomes acceptable to the DOE (48). The perspectives represented by each of the three factors disagreed with this, including the one on which the DOE and Kaiser-Hill representatives loaded significantly. The next highest negative consensus statement was that the process does not make conflict worse (38). Clearly, the desire to address the RSAL issue overrides any tendencies these participants may have for avoiding conflict.

Distinguishing Statements

There are important areas of differentiation among the perspectives. Table 3 identifies the three most important themes, in descending order, in each of the perspectives whose narratives were presented above. They do not completely share the same three themes. Moreover, those that are shared are not weighted equally.

One important area of disagreement involves the use of consensus and the need to address power imbalances. Perspective B places high importance on the use of consensus to protect stakeholders, while perspectives A and C prefer to not use consensus. In perspective A, consensus is viewed as a barrier to making decisions. In perspective C, there is a worry that consensus leads to "watered down" recommendations.

Second, the role of science and technical information is emphasized differently in each of the perspectives represented by the three factors. It is most important in perspective C. As noted above, the most important aspect of this perspective is that the process should focus on scientific and public health aspects when setting soil action levels for the site. On the other hand, perspective B places much less importance on science and technical information. To be sure these are important, but according to those who loaded highly on factor B so are other kinds of information. Finally, perspective A

Table 3. Top Three Themes Emphasized in Each Perspective.

Perspective A	Perspective B	Perspective C
Responsible collaboration among key groups	Maximize effective widespread participation	Sound science and information
Sound science and information	Minimize power imbalances, rely on consensus	Having the right attitude about participation
Produce feasible outcomes	Low importance on science and information	Authorities make decisions, do not use consensus

emphasizes the importance of scientific information in planning and decision-making but de-emphasizes the importance of broad access to that information. The distinction between the importance of *types* information and *access* to that information is further suggested by the next area of disagreement.

Third, there is a divergence about access and who should be a participant. Both perspectives A and C suggest limitations on who can participate. Allowing restrictions on who can participate (12) was ranked highest in perspective A – only "important" stakeholders should be included. According to perspective C, there is a need for participants to have technical understandings of the issues, because complex issues should not be oversimplified. Both of these perspectives also give higher support for participants having reasonable expectations about what the DOE can do (15). In contrast, perspective B places strong emphasis on maximizing widespread participation by promoting access and openness. In this perspective, attention should be given to making meetings and information broadly accessible. Restrictions on who can participate can marginalize important viewpoints.

Fourth, while each perspective places importance on the character of relationships and interactions among participants, they are concerned with different aspects. In perspective A, participants, including the DOE, should be accountable and honest; they should see beyond their self-interests to the larger good (11), they should attend meetings regularly (17), and the DOE should respond to requests for information, questions, etc. in a timely manner (24). Accountability and a concern that participants adopt an orientation toward the common-good (11) are also apparent in perspective C. In addition, perspective C emphasizes trust among the participants. In perspective B, the weaker emphasis given to statement 11 reflects the concern that the viewpoints of some not be marginalized because more powerful stakeholders define them as contrary to the "common good." In perspective B, the emphasis is placed on ensuring that participants feel listened to (1).

The Links between Remediation and Long-Term Stewardship

As illustrated by the similarities and differences among the factors and their associated perspectives on process for remediation and stewardship decision-making, there are different preferences regarding features for the most appropriate process to establish soil action levels for Rocky Flats. In addition, differences among each of the perspectives emerging from the analysis suggest that they place different emphases on the need to integrate remediation and stewardship planning. Perspective C reflects most strongly the traditional approach of DOE that remediation and stewardship planning can be separated, and, in fact, individuals from DOE and Kaiser-Hill who participated in our study loaded significantly on factor C. According to perspective C, the process should be efficient and focused to ensure that soil action levels are established in a timely manner so that site closure would be achieved on schedule. Soil remediation is the focus and there is little concern at this stage on stewardship planning. On the other hand, those subscribing to the perspectives represented by factors A and B have a stronger desire to link remediation and stewardship planning. This is most apparent in perspective B, suggested by both reluctance to be tied to DOE's agenda or schedule for decisions and concern with how near-term remediation decisions may affect long-term stewardship. All of the individuals loading significantly on these factors were members or staff of the RFCAB or RFCLOG. Both these groups have expressed their concern that DOE consider remediation decisions (such as setting of RSALs) in the broader context of long-term stewardship planning.

Limitations of this Study

There are two issues we wish to highlight that limit the findings from this study. First, while this study revealed three distinct preferences for process design among the 12 people who participated in our research, we cannot claim that these are the only perspectives that exist among the full spectrum of stakeholders involved in Rocky Flats remediation and stewardship planning. A larger study may indeed reveal more perspectives. However, the performance of this research is defended by having research instruments designed with the experience of several other similar applications and by having assurance, through our conversations with local stakeholders, that we selected people with very different yet well-formed opinions about process. In addition, an inherent limitation of Q methodology is that we cannot make any claims about the prevalence of these perspectives in the larger population of people involved with this case study. A survey could be used to find what kinds of people subscribe to particular perspectives, but that topic was not part of this research.

Second, we did not ask people specifically about their views on linking remediation and stewardship planning. Instead, we allowed concerns about stewardship to be expressed through the way they sorted the statements on process and on their reflections about the sort (and recorded on audio tapes). Thus, our ability to fully understand how and why people link these activities is limited.

CONCLUSION

A persistent challenge for agencies responsible for protecting communities from long-lived contamination is how to construct a decision-making process that meets the needs and expectations of all involved, while producing implementable and effective policy outcomes in an efficient manner. This refers to the question of how best to match method with purpose. Scores of reports advise how best to conduct a decision-making process that links good science with good participation – or analysis with deliberation, in the words of the National Research Council report *Understanding Risk* (1996). Many emphasize that methods of incorporating public participation into risk-related decision-making must be tailored to the situation at hand (Chess & Purcell, 1999; National Research Council, 1996, 2000). Even so, scholars and practitioners of public participation in environmental and risk-related decision-making have often presumed that specific features can be used to define "best practices." For example, the National Research Council (2000, 2003) has proposed a number of characteristics that should guide remediation and stewardship planning processes at DOE facilities.

In our research we have asked whether or not, for a given situation, there is a single point of view, or perspective, of the best way to conduct environmental and risk decision-making with the public in the long term. In the case of setting soil action levels for plutonium at Rocky Flats, the answer is clearly "no." In our study we found three distinct, sometimes even competing, perspectives about what would be the most appropriate process. The existence of multiple perspectives is a result that has been replicated in a number of other case studies we have conducted about environmental health, forest management, remediation of radiological contamination, radiation health effects studies, and watershed and coastal management (Webler, Tuler, & Krueger, 2001; Webler & Tuler, 2001; Tuler, Webler, & Finson, 2005; Webler, Tuler, & Tanguay, 2004). Even in the same context, we find that different stakeholders have different ideas about what makes for a meaningful and successful decision-making process.

In addition, two of the perspectives emerging from our analysis suggest the importance placed by stakeholders on linking contaminant reduction, isolation of remaining contaminants, and long-term stewardship activities – a view that stands in contrast to DOE's approach (National Research Council, 2003) which is more closely represented by the third perspective. The National Research Council suggests two broad reasons for linking remediation and stewardship planning. First, remediation decisions can affect the characteristics of long-term institutional management systems that are needed. Second, the processes of planning and conducting remediation create their own context in which future stewardship planning efforts are embedded. How planning and remediation are accomplished and the degree to which people are satisfied with the process can have important implications for the confidence placed in DOE and other stewards to meet their responsibilities. Recent actions to dissolve the RFCAB and RFCLOG and establish a new local advisory group have underscored this dynamic. Disagreements about process that we found reflected in the three perspectives are now being played out again as the DOE, local governments, and others strive to craft a new process for stewardship-related activities and implement the strategic plan of the DOE's Office of Legacy Management (DOE, 2004).

The findings from this project have important implications for government agencies that wish to improve their efforts to involve the public in planning and other activities related to the management of environmental health and the planning for long-term stewardship of contaminated lands. First, planners should assume that there are many ideas about what is the most appropriate process for a given situation. The challenge for planners is to clarify points of consensus and dissensus and promote a dialogue about what would be the most appropriate process. We echo others' cries about the importance of early involvement among interested and affected parties in the design of the process (National Research Council, 1996). Process design should be a deliberative activity. At the same time, planners must attend to constraints about, for example, budget and DOE's regulatory responsibilities.

Second, success is a function not only of the design features used, but also the extent to which the design matches the needs and preferences of the participants, including the responsible federal and state agencies, *in a particular context*. If planners do not address conflicting expectations about what is an appropriate process, then even the best-run process may not be deemed successful or legitimate by important stakeholders (Ashford & Rest, 1999). At the same time, it might not be important to include every possible feature needed for best practices in every case. Our findings suggest that some features may not be important to participants because of the nature of the particular problem being faced and the particular context of the process. Organizers may take heart in knowing that assessment and decision-making processes can still be broadly seen as meaningful and successful even if they do not manifest every attribute of good practice, *provided they meet the most salient demands of the context*. This is also why organizers cannot assume that a process successfully implemented in one context will be appropriate in another.

In other words, those responsible for organizing risk assessment and decision-making processes, and those participating as well, would improve their chances of having a meaningful and successful experience if they take the time to think, engage in deliberation, and strive for agreement about what is the most appropriate process for a given instance. Paralleling calls for flexibility, accountability, and learning in management institutions for long-term stewardship, crafting a decision-making process that engenders learning and is flexible enough to change as the need arises and in response to the will of those served should become the overriding concern of all involved in process design and implementation.

NOTES

1. The RFCAB, one of eleven of the Department of Energy Environmental Management's site-specific advisory boards, includes members from diverse backgrounds and stakeholder groups. Members represent pre-defined constituencies, including academic institutions, business, local communities, local governments, health care providers, public interest groups, environmental organizations, Rocky Flats employees, and technical disciplines.
2. Town of Westminster and City of Broomfield are two communities adjacent to the Rocky Flats site.
3. The Coalition of Local Governments was established to provide a way for local governments to work collaboratively on issues relating to remediation of the site.
4. One measure by which we are confident that our statements are broadly representative is that after each Q sort we asked if the subject could think of any additional elements that would be important to him or her that were not represented in the Q statements. None of the 117 people who completed sorts in our 10 case studies in our full research project suggested additional elements.
5. A question has arisen among researchers using Q methodology about whether the pattern into which people are required to sort the Q statements matters to the results that are obtained. The conclusion among researchers of Q is that the use of a normal distribution makes little or no difference to the results of a study. We elected to use the normal distribution because we find it helps people sort the cards and because it enables us to use software that we prefer.

ACKNOWLEDGEMENTS

We thank the people who agreed to participate in our research and Judith Bradbury for her insights about public involvement at Rocky Flats. We also thank Will Focht of Oklahoma State University for his advice on Q methodology. Finally, the astute comments of the reviewers were also helpful in improving earlier drafts. This paper is based on work supported by the National Science Foundation under grant 0114784. Any opinions, findings, and conclusions or recommendations expressed in this material are those of the authors and do not necessarily reflect those of the National Science Foundation.

REFERENCES

Applegate, J., & Sarno, D. (1997). FUTURESITE: An environmental remediation game. *Simulation and Gaming*, *28*(1), 13–27.

Ashford, N., & Rest, K. (1999). *Public participation in contaminated communities*. Cambridge, MA: Center for Technology, Policy, and Industrial Development, Massachusetts Institute of Technology.

Beierle, T., & Cayford, J. (2002). *Democracy in practice: Public participation in environmental decisions*. Washington, DC: Resources for the Future.

Bradbury, J., Branch, K., & Malone, E. (2003). *An evaluation of DOE-EM public participation programs*. Report prepared for the DOE Environmental Management Program. Report PNNL-14200. Washington, DC: Pacific Northwest National Laboratory.

Brown, S., (1986). Q technique and method: Principles and procedures, In:, W.D. Berry, & M.S. Lewis-Beck (Eds), *New tools for social scientists*. Thousand Oaks, CA: Sage

Brown, S. (1996). Q methodology and qualitative research. *Qualitative Health Research*, *6*(4), 561–567 Available at www.rz.unibw-uenchen.de/~p41bsmk/qmethod/srbqhe.htm.

Chess, C., & Purcell, K. (1999). Public participation and the environment: Do we know what works?. *Environmental Science and Technology*, *33*(16), 2685–2692.

Department of Energy, US Environmental Protection Agency, and Colorado Department of Public Health and the Environment. (1996). *Action levels for radionuclides in soils for the Rocky Flats cleanup agreement*. Final Report. Washington, DC: Department of Energy.

Department of Energy (DOE) (1999). *From clean-up to stewardship*. Washington, DC: Office of Environmental Management, Department of Energy.

Department of Energy. (2001a). *A report to Congress on long-term stewardship. Volume 1: Summary report*. Washington, DC: Office of Environmental Management, Department of Energy.

Department of Energy. (2001b). *A report to Congress on long-term stewardship, Volume 2: Rocky Flats Environmental Technology Site* (pp. 85–105). Washington, DC: Department of Energy.

Department of Energy. (2003). Public participation and community relations. Policy document DOE-P-141.2, approved 2 May 2003. Washington, DC: Office of Congressional and Intergovernmental Affairs, Department of Energy.

Department of Energy (2004). *Office of legacy management strategic action plan*. Washington, DC: Department of Energy Office of Legacy Management.

Drew, C., Grace, D., Silbernagel, S., Hemmings, E., Smith, A., Griffith, W., Takaro, T., & Faustman, E. (2003). Nuclear waste transportation: Case studies of identifying stakeholder risk information needs. *Environmental Health Perspectives*, *111*(3), 263–272.

Earle, T. (2004). The Rocky Flats controversy on radionuclide soil action levels. In: *Stakeholder participation in radiological decision making: Processes and implications*. Case studies for the 3rd Villigen Workshop, Villigen, Switzerland, 21–23 October 2003 (pp. 21–32). Paris: Nuclear Energy Agency, Organization for Economic Cooperation and Development.

Environmental Law Institute (2001). *The role of local governments in long-term stewardship at DOE facilities*. Washington, DC: Environmental Law Institute.

Environmental Protection Agency. (1996). *Final report of the federal facilities environmental restoration dialogue committee: Consensus principles and recommendations for improving federal facilities cleanup*. Federal Facilities Restoration and Reuse Office. Washington, DC: Environmental Protection Agency. Available at http://www.epa.gov/swerffrr/comminvolve/fferdc.htm.

Environmental Protection Agency. (2003). Public involvement policy of the US Environmental Protection Agency. EPA-233-B-03-002. Available at www.epa.gov/policy2003.

Fernald Citizens Task Force (1995). *Recommendations on remediation levels, waste disposition, priorities, and future use*. Ross, OH: The Fernald Citizens Task Force.

Focht, W. J. (1995). A *heuristic political inquiry into NIMBY conflict: Exploring solutions to gridlock*. Unpublished Ph.D. dissertation, Oklahoma State University, Oklahoma.

Institute for Energy and Environmental Research. (2001). *Setting clean-up standards to protect future generations: The scientific basis of subsistence farmer scenario and its application to the estimation of radionuclide soil action levels (RSALs) for Rocky Flats*. Takoma Park, MD: Institute for Energy and Environmental Research.

Ledwidge, L., Moore, L., & Crawford, L. (2004). Stakeholder perspectives on radiation protection. *Health Physics*, *87*(3), 293–299.

McKeown, B., & Thomas, D. (1988). *Q methodology*. Sage University Paper Series on Quantitative Applications in the Social Sciences 07-066. Beverly Hills, CA: Sage.

National Research Council (NRC) (1996). *Understanding risk: Informing decisions in a democratic society*. Washington, DC: National Academy Press.

National Research Council (2000). *Long-term institutional management of US Department of Energy legacy waste sites*. Washington, DC: National Academy Press.

National Research Council (2003). *Long-term stewardship of DOE legacy waste sites – a status report*. Washington, DC: National Academy Press.

Risk Assessment Corporation (RAC). (2000). *Final report to the radionuclide soil action level oversight panel*. RAC Report No. 1-RSALOP-RSAL-2000-FINAL. Neeses, SC: Risk Assessment Corporation.

Risk Assessment Corporation (RAC). (1999). *Final report task 1: Cleanup levels at other sites*. Report to the Radionuclide Soil Action Level Oversight Panel. RAC Report No. 6-RSALOP-RSAL-1999-FINAL. Neeses, SC: Risk Assessment Corporation.

Rocky Flats Citizens Advisory Board. (2001). *Stakeholder workshop on computer modeling and parameter selection for radionuclide soil action levels at Rocky Flats*. Summary of workshop 27–28 April 2001. Westminster, CO: Rocky Flats Citizens Advisory Board. Available at www.rfcab.org.

Rocky Flats Stewardship Working Group. (2001). *Hand-in-hand: Stewardship and cleanup*. Report to the Rocky Flats Coalition of Local Governments and the Rocky Flats Citizens Advisory Board. Westminster, CO: Rocky Flats Citizens Advisory Board. Available at www.rfcab.org.

Rood, A., Grogan, H., & Till, J. (2002). A model for a comprehensive assessment of exposure and lifetime cancer incidence from plutonium released from the Rocky Flats Plant, 1953–1989. *Health Physics*, *82*(2), 182–212.

Till, J., & Meyer, K. (2001). Public involvement in science and decision making. *Health Physics*, *80*(4), 370–378.

Tuler, S., Webler, T., & Finson, R. (2005). Competing perspectives on public involvement: Planning for risk characterization and risk communication about radiological contamination from a national laboratory. *Health, Risk, & Society*, *7*(3), 247–266.

Webler, T., & Tuler, S. (2000). Fairness and competence in citizen participation: Reflections from a case study. *Administration and Society*, *32*(5), 566–595.

Webler, T., & Tuler, S. (2001). Public participation in watershed management planning: Views on process from people in the field. *Human Ecology Review*, *8*(2), 29–39.

Webler, T., Tuler, S., & Krueger, R. (2001). What is a good public participation process? Five perspectives from the public. *Environmental Management*, *27*(3), 435–450.

Webler, T., Tuler, S., & Tanguay, J. (2004). Competing perspectives on public participation in National Park Service planning: The Boston Harbor Islands National Park Area. *Journal of Park Administration*, *22*(3), 91–113.

LONG-TERM MANAGEMENT IN PRACTICE: LESSONS FROM THE SITES

LONG-TERM STEWARDSHIP FOR THE HANFORD NUCLEAR SITE

Max S. Power

ABSTRACT

The Hanford site, in Washington State, presents a large-scale test for long-term stewardship of residual radioactive and chemical contamination. Large quantities of waste and contaminated materials will remain in perpetuity near the land's surface and the Columbia River, making Hanford perhaps the most complex long-term stewardship challenge among former weapons production sites. This paper explores the scope of contamination, the Department of Energy's approach to cleanup, the likely residuals requiring long-term stewardship, and the issues that are thus raised. Expectations with respect to long-term land uses, the likely durability of institutional controls, and funding, information management, and accountability have long been influenced by mistrust and tension between local communities, regulators, tribes, and the Department of Energy (DOE). Despite positive steps by DOE such as the creation of the Office of Legacy Management, DOE's dependence upon annual appropriations, its sovereign immunity with respect to key state and local land use regulations, and recent policy pronouncements that cast doubt on the willingness to respond to unanticipated problems with additional cleanup, all leave doubt in the minds of regulators and communities that DOE can be counted upon to be both proactive and accessible once cleanup is complete.

Long-Term Management of Contaminated Sites
Research in Social Problems and Public Policy, Volume 13, 81–104

ISSN: 0196-1152/doi:10.1016/S0196-1152(06)13003-3

INTRODUCTION: HANFORD'S LEGACY OF CONTAMINATION

The Hanford site, in Washington State, presents a large-scale test for long-term stewardship of residual contamination. Hanford's size, the complexity of previous operations, and the large quantities of waste and contaminated materials that will remain in perpetuity near the land's surface make both active "cleanup" at Hanford and long-term protection of public health and the environment extremely challenging. This paper explores the scope of contamination, the approach to cleanup, the likely residuals requiring long-term stewardship, and significant issues about how that stewardship can be assured.

The Hanford site, straddling the Columbia River in southeastern Washington State, was selected in 1943 to be the Manhattan Project's manufacturer of plutonium for nuclear weapons. In a truly remarkable engineering and construction feat, the first large-scale nuclear reactor (or "pile") went critical little more than a year after construction started, and the first plutonium was separated from irradiated fuel by the end of 1944.[1] In all, nine large reactors would be built along the Columbia River and five different separations facilities to retrieve plutonium ("canyons") would rise on the plateau in the center of the site. The Manhattan Project and its successor agencies would create hundreds of ancillary facilities, including those to manufacture uranium fuel for the reactors and others to process and stabilize separated plutonium.

By the time Hanford's plutonium production mission ended officially in 1989, Hanford had produced two-thirds of the United States' stockpile of about 100 metric tons of separated plutonium (DOE, 1996). Along the way, Hanford also undertook some nondefense nuclear activities. By the early 1990s Hanford was estimated to have nearly a million cubic meters of stored or disposed wastes, with more than 350 million curies (Ci) of radioactivity, some 23 million cubic meters of contaminated solid media, and 1.2 billion cubic meters of contaminated water (DOE, 1997). Discharges of some 440 billion gallons of process wastewater from various facilities and shallow-land disposal of solid wastes had created about 1,500 identified waste sites.

Three areas on the site will be of primary concern in long-term stewardship (Fig. 1). While final decisions have not been made, the present outlook is that very long-lived and substantial residual contamination will remain in these areas:

- Contaminants released into the soil and/or groundwater in the reactor area (the so-called 100 Area) and the fuel fabrication area (the 300 Area),

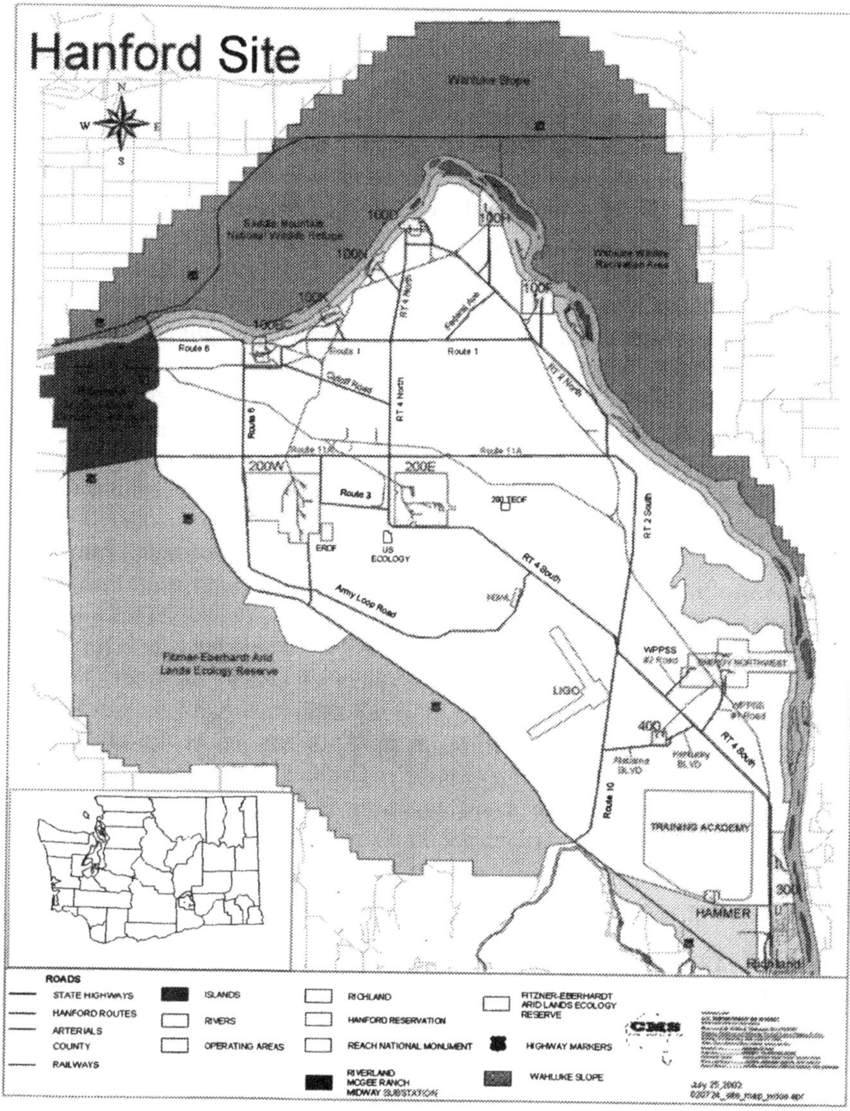

Fig. 1. Hanford Site.

both near the Columbia River. Changes in river levels resulting from seasonal fluctuations and operation of upstream dams, precipitation, and intentional or accidental surface releases of water (e.g., water line leaks) can all drive these contaminants into the River.

- The "tank farms" in the 200 East and 200 West Areas. These locations contain 177 tanks where liquid residuals from chemical separation of plutonium (high-level waste) and the cleaning of processing facilities have been stored. Most of them are quite large, 75 ft in diameter and having a capacity of one million gallons. One hundred and forty-nine of the tanks are "single shell"– a single, carbon steel tank set in a concrete vault. Of these, 67 are known or suspected to have leaked more than one million gallons of highly contaminated wastes to the ground. While current efforts are intended to remove liquids, sludges, and solids from these tanks for treatment and stabilization, it is likely that the tanks themselves, some irretrievable residuals, some of the complex piping and transfer equipment that link tanks to one another and to processing plants, and contaminated soils around the tanks will all be left in place.
- Shallow-land disposal grounds in the 200 Areas. By the time Hanford is "cleaned up" and disposal facilities close, the "low-level burial grounds" will likely contain more than a million cubic meters of low-level and mixed low-level radioactive and hazardous waste from Hanford generators, from the U.S. Navy, and from other Department of Energy facilities. To date, U.S. DOE has disposed of more than 6.3 million tons of soil and debris from surface cleanups in the 100 and 300 Areas in the Environmental Restoration Disposal Facility (ERDF). The five large "canyon" facilities used to separate plutonium and uranium from spent nuclear fuel, also located in the 200 areas, will either be dismantled and placed in an ERDF-like facility or more likely, be collapsed in place and become de facto waste burial grounds.

In the case of both tank farms and disposal grounds, eventual failure of engineered barriers and breakdown of waste forms, and possibly even climatic change, may increase groundwater contamination, which would eventually reach the Columbia River.

HANFORD'S APPROACH TO CLEANUP

The approach and schedule for remediation and closure of Hanford's contaminated sites is embodied in the Hanford Federal Facility Agreement and

Consent Order (DOE, 2003a), better known as the Tri-Party Agreement (TPA). The agreement was signed in 1989 by the U.S. DOE, U.S. Environmental Protection Agency (EPA), and the Washington State Department of Ecology. The TPA lays out procedures and schedules for bringing Hanford into compliance with both the Resource Conservation and Recovery Act (RCRA), as implemented by the State of Washington, and the Comprehensive Environmental Response, Compensation and Liability Act (CERCLA). In addition, DOE has responsibilities for meeting its own requirements relating to nuclear facilities under the Atomic Energy Act.

Briefly, the Hanford cleanup approach has four prongs:

- *Move from the outside in, and shrink the geographic scope of the problem.* Thus, spots of nonradioactive contamination in the Fitzner-Eberhardt Arid Lands Ecology Reserve, the Wahluke Slope (north side of the Columbia), and the logistical areas adjacent to the City of Richland (1100 Area) shown in Fig. 1 were cleaned up, and those areas removed from the CERCLA Superfund list. Work to remove contaminated buildings, contaminated soils (to a depth of 15 ft or more, where structures or significant contaminants are found), and wastes disposed in the 100 and 300 Areas near the Columbia River is more than halfway complete. USDOE expects to complete this work by 2012 in the 100 Areas and 2018 in the 300 Area.
- *Where possible, remove source terms from the environment.* For example, DOE is removing buried wastes from the 100 Areas and highly contaminated soils under a liquid discharge trench in the 100-N Area as well as using pump-and-treat systems to remove hexavalent chromium and carbon tetrachloride, both highly toxic, from both soil and groundwater.
- *Where appropriate, retrieve, treat, and send highest risk wastes to deep geologic disposal.* By far, the largest, most expensive and most risk-reducing source term removal will be retrieval and solidification of the Hanford tank wastes. Once retrieved, the wastes will be separated into high-and low-activity fractions, and the high-activity fraction will be prepared for disposal in a geologic repository. Transuranic wastes segregated in shallow trenches from 1970 onward are being retrieved, processed, and sent to the Waste Isolation Pilot Plant (WIPP) in New Mexico. If older wastes containing sufficient transuranic elements are exhumed as part of the cleanup process, they too will go to WIPP.
- *Protect remaining contamination and disposed wastes with adequate engineered, institutional, and administrative controls.* The planning assumptions for the cleanup are: contaminated soil and debris from the river corridor are disposed in the ERDF, which is lined and has a leachate

collection system; permits are required for excavation and drilling on the Hanford site; contaminated areas are fenced and/or posted; there is a central land use review process for the site; and the ERDF and low-level waste burial grounds will be capped to prevent water infiltration and plant and animal intrusion.

Projected schedules for completing these activities vary. In its recent Performance Management Plan, DOE proposes to complete all cleanup activities at Hanford, including closure of waste disposal facilities, by 2035 (DOE, 2002a). Cleanup activities in the 100 and 300 Areas would be completed in 2012 and 2018, respectively. Remedial action and facility deactivation in the center of the site would be completed by 2021, and tank waste retrieval and treatment by 2028 (the original milestone in the TPA). All spent nuclear fuel, high-level and transuranic wastes would be shipped off-site by 2033.

"End States": Where does DOE's Long-Term Stewardship Begin?

For purposes of this paper, long-term stewardship is defined by seven tasks, as proposed by the National Research Council's Committee on Long-Term Institutional Management of DOE Legacy Waste Sites (NRC, 2003): guardian, watchman, land manager, repairer, archivist, educator, and trustee. These tasks, as they apply to the Hanford site, will be discussed in more detail below.

Exploration of long-term stewardship for Hanford and other DOE nuclear sites consistently reveals a tension between state regulators, local communities, affected tribes, and other stakeholders, on the one hand, and the DOE on the other. In varying degrees, the stakeholders press for cleanup for "unrestricted use." In other words, *they seek a state of affairs where long-term stewardship is not needed.*

There are several reasons for this desire for "complete" cleanup and avoidance of the need for long-term stewardship. Three are mentioned most often, at least at Hanford, but they boil down, in the end, to a lack of trust and confidence that the federal agency involved will act in a truly protective way, as opposed to minimizing near-term costs.

- The local community must rely on institutional controls in the long term to protect its citizens and environment. But institutional controls have not proven reliable,[2] and to make matters worse, DOE's position has been that it cannot be bound by state and local land use controls (Miller, 2002).

- Monitoring, maintenance, repair of physical structures, records management, and emergency response in the event of the failure of containment or isolation must be funded out of annual appropriations. DOE is not subject to financial assurance requirements under RCRA and CERCLA, and has taken the position that it cannot set aside funds to be expended in the future.[3] One consequence is that long-term costs may not be factored into remedial decisions. Another often-heard concern is a lack of confidence in the long-term protectiveness of physical as well as institutional controls (NRC, 2000).
- At Hanford, and some other sites, American Indian Tribes have reserved treaty rights to resources. The tribes often assert that these rights have precedence over remedial decisions that leave in place restrictions on or prohibitions to the exercise of those rights. Moreover, the tribes argue that they are disproportionally affected by the potential for future contamination via environmental pathways, due to cultural factors, such as diet, use of natural medicines, and spiritual or religious practices (STGWG, 2001; Harris, 1998).

The DOE's view has been that cleanup to unrestricted use, even if technically feasible, does not focus resources on highest risk reduction and may distort priorities. The "Top-to-Bottom Review Team" chartered by then-Secretary of Energy Spencer Abraham delineated this position in its 2002 report to then-Assistant Secretary for Environmental Management Jessie Roberson:

> Cleanup of the sites is often further complicated by a lack of realistic future land-use assumptions, and by scenarios that assume that highly contaminated areas will be subject to farming, drilling of wells, or residential use .... Another major factor affecting USDOE cleanups is point of compliance for groundwater contamination. To the extent that the points of expected compliance with state and U.S. Environmental Protection Agency (EPA) standards are located near areas unlikely ever to be released for public use, unrealistic goals for cleanup are established.
>
> The overriding goal is to reduce risk and protect public health and safety. To accomplish that goal, priorities and the associated funding needs should be established at Headquarters after consultation with the sites, and should be based on the process of risk reduction, not merely controlling risk (DOE, 2002b).

To the untrusting, this statement means (a) that DOE believes it need not comply with environmental standards, (b) that it will control future land uses, and (c) that field – i.e., local – commitments and agreements will be overridden by headquarters. Suggesting that land is "unlikely ever to be released for public use" as reason to avoid compliance with environmental standards implies a degree of certainty about long-term future land use that many believe unjustified. As will be discussed later, that is particularly true when "continued federal ownership" is proposed by USDOE as sufficient to demonstrate the way it will address the seven tasks of long-term stewardship.

Some of the distrust reflects the earlier response by acting Assistant Secretary for Environmental Management James M. Owendoff to state and tribal concerns about the Department's intent to rely on long-term controls to leave contamination in place:

> The overall goal of DOE's cleanup program is to design cleanup strategies that protect human health and the environment. The level of cleanup is based on the expected future land use. We believe that if, after a site is remediated to levels appropriate for the specified land use, communities decide that they desire further cleanup to allow for less restrictive land uses, then the cost of such additional cleanup should not be borne by the Department of Energy (Owendoff, 1999).

To skeptical local government officials, this suggests that DOE will rely on its own ideas about future land use to limit its liability, while thrusting the costs of future control and/or land use changes onto the local community.

It is not surprising, then, that much of the discussion about long-term stewardship has revolved around expected land uses and mechanisms for controlling or restricting land uses based on residual contamination, i.e., on institutional controls. These are the areas where local and state governments believe they have both more experience and better legal mechanisms than does DOE.

At Hanford, the effort to define expected land uses was launched in 1991. DOE, EPA, and the Washington Department of Ecology jointly convened the Hanford Future Site Uses Working Group (Hanford FSUWG). The group represented local governments, affected Tribes, and regional labor, business, agricultural, and environmental interests. The working group's report had significant implications for long-term stewardship at Hanford for three reasons:

- It set the pattern for dealing differently with different portions of the site.
- The report also recognized that different timeframes would apply to surface and groundwater cleanup as well as to cleanup in different areas.
- Its projected land uses were based on a broad consideration of public values for the use of different parts of the site. The FSUWG presented a range of likely and desirable land uses to guide the level of remediation for the different parts of the site. The working group did not analyze residual risk. Their list of desirable land uses (Hanford FSUWG, 1992) formed the basis for the alternatives examined in the Hanford Comprehensive Land Use Plan (DOE, 1999).

The division of the Hanford site into six geographic zones (Fig. 2) led to recommendations to accelerate cleanup of nonradiological contamination in the large buffer areas (Arid Lands Ecology Reserve and Wahluke Slope), even though these areas were relatively "low risk." It also highlighted the potential for highly valued uses along the relatively unspoiled Columbia River shoreline outside the reactor (100) and fuel fabrication and laboratory (300) areas. The river corridor outside the 100 and 300 Areas was subsequently included in the Hanford Reach National Monument by Presidential Proclamation (Federal Register, 2000), and will require minimal long-term stewardship attention.

For the 100 Areas, the Future Site Uses report envisioned a range of uses that would be compatible with unrestricted *surface* use. It also set unrestricted groundwater use as a goal, but acknowledged that the goal might not be reached for a long time. The stakeholder group wanted to "get on with cleanup" to support other high-value river-related uses, while recognizing that technology for groundwater cleanup might not be presently available. In subsequent regulatory agreements, the achievement of unrestricted surface use was measured by use of a "residential farmer" scenario (EPA, 1995).

This has been a source of continuing confusion, because it does not necessarily imply the expected land use. The residential farmer scenario, however, appears to balance the risk requirements of CERCLA guidance, Washington State's Model Toxics Control Act, and the exercise of Indian treaty-reserved rights.

While the FSUWG was developing its report, DOE made a decision to "safe-store" eight of the nine reactor cores for 75 years, to allow radioactive decay before workers would tackle their removal (DOE, 1991). The FSUWG acknowledged that small areas containing the reactor buildings themselves would remain fenced and restricted from public use during that period.

Thus, institutional controls restricting excavation and groundwater use will apply in the 100 Areas for a considerable time (varying depending on the particular situation) (cf. DOE, 2001a). In addition, there are operating groundwater pump-and-treat or permeable barrier systems that may need to be maintained past 2012 in order to prevent contaminants from entering the Columbia River.

The situation at the 300 Area, at the southeastern corner of Hanford, is regulatorily complex. From the perspective of long-term stewardship, however, the important point is that USDOE has planned to restrict the core of the area to industrial use – in effect to be redeveloped as a "brownfield" for

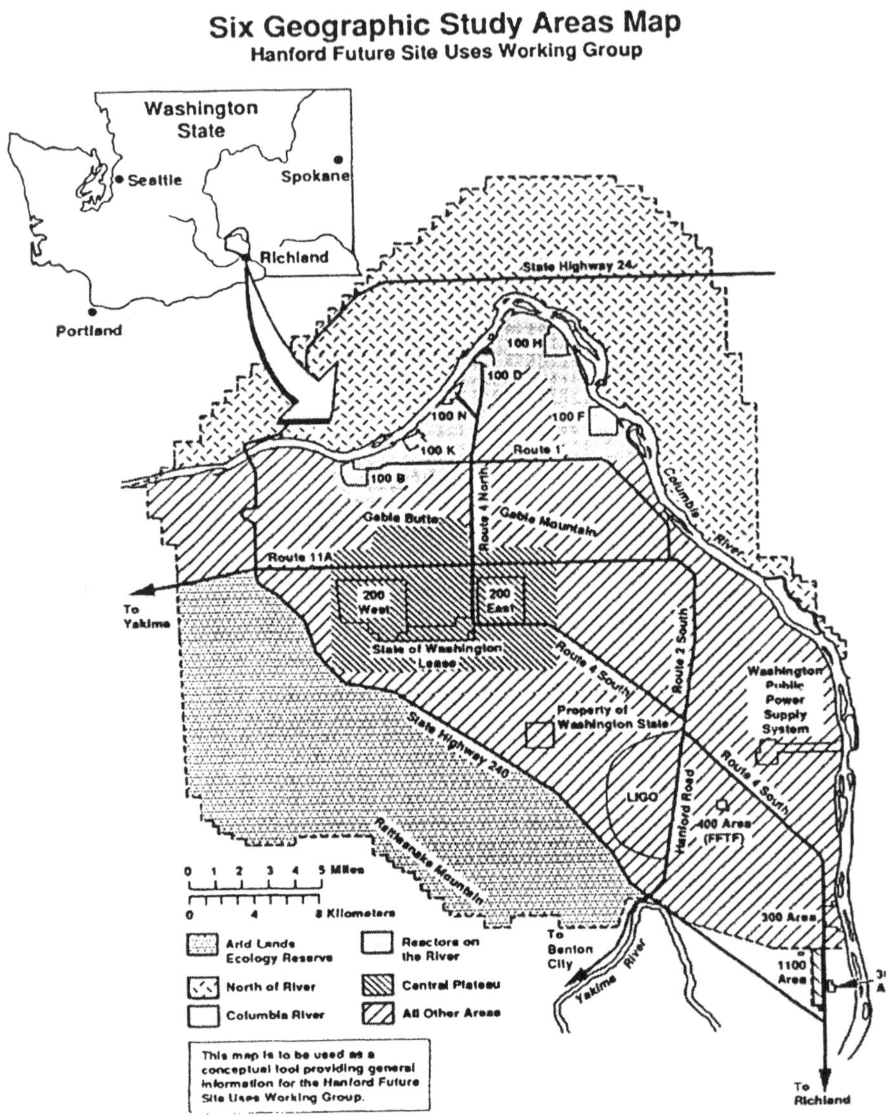

Fig. 2. Designation of Geographic Areas. *Source*: Hanford Future Site Uses Working Group.

new industrial activities. The City of Richland has proposed a mix of uses in its comprehensive plan. The regulatory agencies have questioned whether present CERCLA remedies are sufficiently protective to support the city's plan. If the "brownfield" option is finally selected, it may require occupancy restrictions, perhaps training and orientation for workers within the area, and restrictions on well-drilling and excavation. How such administrative and institutional controls are to be worked out among federal, state, and local agencies has yet to be determined.

In addition, a CERCLA Record of Decision for the groundwater underlying the 300 Area is based on monitored natural attenuation of uranium. It now appears likely that monitoring will indicate a need for more active, long-term remedial measures because the contaminants in the groundwater did not attenuate in the five to ten-year period anticipated in the original ROD.

By far, the most extensive long-term stewardship requirements at Hanford will focus on the "Central Plateau," or 200 Areas and the immediate environs. While many end-state decisions have yet to be made here, the expectation is that significant volumes of waste and contamination will remain at or near the surface in perpetuity. These include DOE's low-level waste burial grounds, the ERDF, a disposal facility for treated low-activity tank waste, and perhaps decommissioned waste tank farms and collapsed chemical processing "canyon" facilities. There is also a commercial low-level radioactive waste site adjacent to the ERDF. In 2001, DOE estimated that 7 million square yards (yd^2) (approximately 2.25 mile^2) of burial grounds and trenches would be capped by 2038 (DOE, 2001a).

The Hanford FSUWG recommended three broad approaches that continue to guide decision-making about cleanup in the central plateau (DOE, 1992, 1999):

- Concentration of wastes from elsewhere at Hanford in the central plateau, while minimizing the use of additional uncontaminated land for disposal. The removal of contamination and buried wastes from the 100 and 300 Areas to the ERDF reflects this precept.
- Prevention of migration of contaminants from these central plateau facilities into the groundwater, and migration of contaminated groundwater, from the 200 Areas toward the Columbia River.
- Creation of a buffer zone around the 200 East and 200 West Areas, where access would be restricted, based on contamination and risk from continuing operations within the "core" 200 Areas. The belief at the time was that this buffer zone could shrink over time, leaving in its wake

unrestricted surface use. While the configuration of the buffer zone has changed, based on information concerning surface contamination, this approach is embodied in current cleanup strategies for the central plateau (DOE, 2002a).

Many of the decisions and actions that will determine the exact nature and extent of remedial actions for the central plateau will not be completed until as late as 2035. The implications of these decisions and actions for long-term stewardship will not be clear for some time. Meanwhile, much of the area is in active use for waste management and treatment, even as cleanup is "complete" in some portions.

Mechanisms for Long-Term Stewardship at Hanford

Faced with the geographic and temporal diversity of the Hanford cleanup, the DOE has developed a strategy for addressing long-term stewardship at Hanford (DOE, 2003b). The "Hanford Long-Term Stewardship Program and Transition" document defines a steward's mission and long-term stewardship functions (DOE, 2003b, p. iii):

> Mission: Provide for continuous human and environmental protection, and the conservation and consideration of use of the biological, natural, and cultural resources, following completion of the cleanup mission.
> Functions:
>
> 1. Manage post-cleanup completion residual risks
> 2. Manage site resources
> 3. Manage stewardship information
> 4. Use science and technology
> 5. Provide post-cleanup completion infrastructure
> 6. Integrate long-term stewardship responsibilities.

DOE's Richland Operations Office developed this program strategy after involving regulators, tribes, and other stakeholders in a series of structured workshops. The document adapts many of the precepts developed in recent years by DOE's Office of Long-Term Stewardship and other national groups (DOE, 2003b). It also generally reflects the seven stewardship tasks identified by the National Research Council Committee and listed above (NRC, 2003).

The Hanford Long-Term Stewardship Program is based on the assumption that areas where the Environmental Management Office's cleanup has been "completed" will be transferred to other entities. In Hanford's case, the most likely recipients are DOE's new Office of Legacy Management, the U.S. Fish and Wildlife Service (as the owner of the Hanford Reach National Monument), and organizations such as the Port of Benton for industrial development of areas adjacent to the City of Richland.

Therefore, the concept of a "transfer checklist" or "turnover criteria" emerges as a key to the strategy (DOE, 2003b). This generally follows the "site acceptance criteria" developed, in draft form, by DOE's Office of Long-Term Stewardship during 2002.[4] This is a structured effort to make sure that all provisions for monitoring, maintenance, information management, institutional and administrative controls, remedy review, emergency response, and liability for residual contamination are clearly identified and assigned to accountable parties.

Developing such a set of criteria is the first of five near-term tasks identified in the Hanford Long-Term Stewardship Program (DOE, 2003b). Using such a site transition framework is also a strategy identified in the draft strategic plan for the new Office of Legacy Management (DOE, 2003c). Given the geographic extent and diversity of cleanup sites at Hanford as well as the differential schedules for remediation in different areas, the Richland Operations Office is attempting to prepare itself for relatively early transition of some parcels. The intent is that the framework, once established, will inform remedial decisions and actions so that "completed" parcels are ready for transfer.

Whether the process will work out so systematically, however, is yet to be seen. The Office of Legacy Management's position is that it will not accept portions of Hanford before the entire site is cleaned up and turned over (Geiser, 2004). As a general policy, the Department of Interior will not accept transfer of lands that have residual contamination requiring any further activity.[5]

In any case, DOE has already begun the process of land transfer at Hanford. One of four designated "Superfund" sites at Hanford – 1100 Area, Arid Lands Ecology Reserve and North Slope – has been deleted from the Superfund site list. These areas contained pockets of nonradioactive hazardous contamination, for which final remedies are in place. A portion of the 1100 Area has been transferred to the Port of Benton. DOE has been negotiating the transfer of the Hanford Reach National Monument, approximately 160–195 mile^2, to the US Fish and Wildlife Service, which presently manages the area. It is not clear whether any specific transfer checklist has been implemented in these cases, or how responsibility for maintaining any necessary institutional and physical controls will ultimately be assigned.

Since 2000, the U.S. EPA has inserted a requirement into Hanford CERCLA records of decision for DOE to prepare a Site-wide Institutional Controls Plan. The effectiveness of the plan is to be assessed annually. To date, the focus of the plan and initial assessments has been near-term. DOE's main concern is whether fences, signs, and administrative controls are in place and have been maintained. So far, DOE has not chosen to address broader questions in the review that would inform land transfer criteria and longer-term performance of institutional controls (Price, 2003).[6]

Issues of Continuing Concern

As cleanup moves forward, perhaps at an accelerated pace, Hanford's regulators and stakeholders will likely argue that the following four long-term stewardship issues must be addressed as remedial action, waste disposal, and closure decisions are made:

- Funding,
- Viability of institutional controls,
- Information management, and
- Accountability.

How each is resolved will depend on a combination of local factors and of national efforts to shape long-term stewardship at contaminated federal sites.

Funding

The communities surrounding Hanford (as well as other DOE nuclear sites) are concerned whether resources will be available in the future to monitor, maintain, and, in case of some failure of a remedy, repair both physical and institutional barriers protecting them from residual contamination – the guardian, watchman, and repairer tasks identified by the National Research Council Committee. Presently, the federal government is exempt from financial assurance requirements under hazardous waste laws, so that funding for these activities in the future will depend on annual appropriations. The communities are understandably concerned that future Congresses will have other priorities and that the communities will bear the cost of protecting public health and the environment. That is, they have relatively little faith in the trustworthiness of the "trustee."

On the other hand, as a federal agency, DOE is bound by the Anti-Deficiency Act. It cannot commit to expend money that has not yet been appropriated, or otherwise bind future Congresses. This area has received a good deal of attention from intergovernmental groups and the Washington, DC-based "think tank" Resources for the Future (Brown, 2002; Bauer & Probst, 2000). It has also been an issue for the US Fish and Wildlife Service at Hanford. Consistent with Department of the Interior policy, the Service asserts that it is not funded to manage areas of past contamination.

Institutional Controls

Institutional controls are generally those legal and communicative barriers that keep individuals from exposure to residual contamination. Examples include deed restrictions against drilling or excavating, land use codes that prevent certain types of development, administrative requirements for access to an area, and signs and fences.

As long as Hanford was a secured area, with controlled access and significant security patrols, institutional controls were not an issue. However, as the security perimeter shrinks so that contaminated areas become relatively accessible to the public, controls become more important. At this point, DOE has procedures in place to control land uses, excavating, drilling, and other activities on the Hanford site. When lands are transferred to other entities, the effectiveness of institutional controls requirements will rely on the diligence of additional parties.

In the early 1990s, DOE entered into an agreement with Benton County that sheriff's deputies would be responsible for dealing with trespass outside of secured portions of the Hanford site. However, land transfer to nonfederal entities may require much more, e.g., deed restrictions or easements to control access, excavation, drilling, and other activities. As noted earlier, a number of studies have pointed to potential problems in establishing, maintaining, and enforcing such restrictions (Miller, 2002; Environmental Law Institute, 2001; NRC, 2000).

Explicit mechanisms to transfer land ownership and responsibility for specific controls are critical. At present, several problems inherent in such transfers suggest a need for layering, redundancy, and relatively frequent reviews of the status of institutional controls. These include:

- The unwillingness of federal agencies to submit to state or local registry or recording requirements;
- The present lack of a mechanism, at least within DOE's field offices at Richland, to systematically assess and integrate institutional controls across both individual operable unit records of decision and the site as a whole, including those portions released to other entities;
- The limited enforceability and low visibility of covenants, deed

restrictions, and easements (as at the infamous Love Canal – see Endnote 2), even if they are agreed to by the present land-owning agency; and
- The cultural pressure to ignore or bury "bad" history about a community that might deter economic development.[7]

DOE's new Office of Legacy Management will build upon the systematic approach developed by the Grand Junction Office of USDOE (now incorporated into Legacy Management) that reviews effectiveness of institutional controls as well as performance of engineered systems at closed uranium mine and mill sites. "Lessons learned" will be applied at larger, more complex sites. However, presenting its early planning, the Office of Legacy Management had regarded institutional controls as secondary to physical barriers as a means to protect public health (Geiser, 2004; Owen, 2003).[8]

Information Management and Preservation

A third area of continuing concern relates to retention, updating and accessibility of information (the archivist and educator tasks). In 2001, DOE's State and Tribal Government Working Group's committee on long-term stewardship surveyed states and tribes hosting or adjacent to USDOE nuclear facilities (STGWG, 2001). Respondents identified several kinds of information that should be maintained as long as long-term stewardship is required. The four kinds of information most often listed as necessary are location and inventory of contamination and wastes, physical controls, institutional controls, and accountability.

In this case, "accountability" includes knowing who is responsible for gathering and reporting information on whether a remedy is performing as expected, whether contaminants are migrating, whether physical maintenance has been addressed, and whether institutional controls are being effectively enforced. Survey respondents also identified the need to have this information available to respond to emergencies such as floods, earthquakes, and wildfires.

Availability, accessibility, and currency are all concerns with regard to stewardship information. This will be particularly true at a site where there may be no continuing presence of the responsible party. At Hanford, DOE's departure is presumably 30 or more years off; however, DOE's imminent departure has raised significant questions at sites such as Weldon Spring, Missouri, and Fernald, Ohio.

Will records be available locally? If so, how will they be updated? Will they be in a form that is accessible in future years? This last question reflects concerns about the rapid evolution, and therefore eventual obsolescence of, electronic data systems. If information is made available via the Internet, for example, how will it be translated into current formats so that it remains accessible? At the same time, limited experience with hard-copy administrative records suggests that key identifiers, like those used to identify specific facilities or units, change over time, making tracking very difficult. Present record retention schedules do not necessarily discriminate between records that need to be kept for the long term and those that may be disposed of within a shorter period.

Finally, maintaining *public* accessibility is important to assuring accountability. People in the surrounding community, potential buyers or lessees, recreational users, traditional tribal users – this list is quite extensive – all need to know about potential hazards and to know who is accountable to maintain protectiveness from those hazards. Again, this requires a commitment of resources, not only to provide access but also to make future generations aware that the information is available (part of the "educator" function identified by the NRC committee).

The author and a colleague have suggested two possible mechanisms to provide future access and awareness (Price & Power, 2002). The first creates trustees through the creation of a trust to oversee monitoring, maintenance, and information management activities. There would be one or more persons who have a legally enforceable fiduciary responsibility to make sure that these activities are appropriately executed. The trust focuses accountability.

Via the second mechanism, proven institutional forms can become vehicles for intergenerational information transfer. At Hanford, two opportunities present themselves: the creation of the Hanford Reach National Monument, which will have a visitor center, and the proposed "B Reactor Museum." Either could become a repository for records of contamination and cleanup and also provide interpretive exhibits explaining the activities that led to contamination and to cleanup. Either could be assisted by a one-time endowment from the responsible party for the express purpose of creating a stewardship function.

Two enhancements merit consideration in Hanford's case. First, the national monument could develop a relationship with affected tribes at Hanford, including employment of Indian rangers and curators. The tribes are relatively permanent residents of the immediate vicinity and also have strong oral history traditions. Second, either the Reach visitor center or the B-Reactor Museum might establish – again with some initial financial support – a center for research on the history of the Manhattan Project or significant aspects of it, that would draw scholars who would continually mine and reinterpret for coming generations the information resources available.

Accountability
Accountability continues to be an issue for states and communities with DOE cleanup sites in relation to institutional controls, funding and information management as discussed above. The general concern is that the responsible agency shrouded in sovereign immunity and dependent on annual appropriations, will not be accessible and pro-active in protecting the community from residual contamination in the long run.

A related concern is that the authority making cleanup decisions is not accountable for the long-term costs of those decisions. In the minds of many, the 1999 Owendoff letter, referred to above, clearly stated the policy that "you only get one cleanup. After that, you're on your own." The creation of DOE's Office of Legacy Management, while hailed as an important step to address accountability in terms of the "there's nobody there" issue, heightened this second concern.

Governors Gary Locke (Washington) and Bill Owens (Colorado), co-leads for the National Governors Association Federal Facilities Task Force, wrote Michael Owen, the first Director of the Office of Legacy Management to express the concern that Legacy Management would be saddled with cleanups that unnecessarily increased its burden:

> [T]he states must be assured that long-term stewardship concerns are fully integrated into remedial action decisions as they are made. The Office of Legacy Management must not passively "accept" sites for stewardship; it must ensure that the full range of stewardship costs, institutional controls, and operational concerns are considered as part of the cleanup decision process (Locke & Owens, 2003).

At Hanford, this concern was exacerbated by proposals in the DOE's draft "Risk-Based End State Vision" document (DOE, 2004a). This document assumed that the projected land uses at Hanford will be defined by the Hanford Comprehensive Land Use Plan (DOE, 1992, 1999), originally adopted for a 50-year period, for the foreseeable future. The long-term assumption of restricted land use could then be used as a basis to back away from a number of aggressive cleanup actions committed to earlier by DOE. These "variances" include a reduction in retrieval, treatment, and disposal of tank wastes, leaving a number of 100-Area burial grounds in place rather than excavating and moving the contamination to the ERDF on the central plateau, and/or a reduction in removal and concentration of contamination in newly engineered facilities within the central plateau (DOE, 2004a).

As mentioned earlier, a 50-year land use plan and an executive order creating a national monument – both assuming continued federal ownership and control – do not give regulators, local officials, and others a high degree of comfort that long-term land uses will not vary from those envisioned in the risk-based end state vision document. And yet these are assumed to reduce the need for cleanup because other uses need not be contemplated.[9] Experience suggests that population pressures and increasing demand for resources may well overwhelm DOE's near-term end state vision, while residual contamination continues to pose real threats.[10]

In response to stakeholder and regulator concerns, the final Hanford Risk-Based End State Vision calls for unrestricted use in the 100 Areas after 50 years, and supports the approach incorporated in earlier RODs (DOE, 2005). However, as noted above, the central plateau area, including the 200 Areas, will, according to this end-state vision, require active long-term stewardship for hundreds of years.

Regulators and stakeholders see little evidence that DOE has thought through the long-term stewardship implications – the requirements and costs to perform the seven tasks identified above – when proposing such end-states. The creation of the Office of Legacy Management is seen by some as a step to demonstrate accountability for long-term management of risks left in place. Nevertheless, the fear remains that, at Hanford, near-term cleanup decisions are driven by efforts to reduce costs and to complete cleanup projects. Legacy Management has little say in what obligations it will inherit, perhaps 30 years hence.[11]

CONCLUSION

The cleanup of the Hanford nuclear site is a massive, multidecade effort. Only in the past few years has DOE begun to address what long-term stewardship means at Hanford. Overall, Hanford's "closure" is many years away, well beyond imminent closures at such sites as Weldon Spring, Fernald, Mound, and Rocky Flats. However, issues of institutional controls, funding, information management, and accountability are very much on the minds of Hanford's regulators and surrounding communities.

These parties' concerns are heightened by several developments:

- The acceleration of cleanup plans at Hanford, that could result in "closure" along the Columbia River as early as 2012, suggesting long-term stewardship issues need to be addressed sooner rather than later;
- The perceived disconnect between the Office of Environmental Management's drive to a (generally less protective) "risk-based end-state vision" for Hanford and the longer-term requirements of "legacy management"; and

- A number of related actions on the part of DOE that suggest the legacy of contamination left at Hanford and requiring long-term stewardship are greater than anticipated. DOE recently issued a record of decision for the Hanford Solid Waste EIS that envisions importing up to 82,000 cubic meters of additional low-level and mixed low-level radioactive waste for disposal at Hanford (DOE, 2004b).[12] The DOE has also approached Congress with proposals to increase its discretion to leave tank waste in place at Hanford (and other sites) rather than retrieving and sending the waste to a deep geologic repository (Weapons Complex Monitor, 2004).

Skeptical regulators, stakeholders, and tribes are not convinced that the long-term implications – the costs and trustee activities – of these decisions have been addressed by DOE. The level of trust engendered in the federal government's commitment to and capability to perform long-term stewardship will affect how Washington's political leadership deals with proposals by DOE to dispose of wastes from other sites at Hanford (Hoffman, 2004)[13] and with proposed variances based on "risk-based end states."

All of these parties, absent clear and specific commitments to address the seven stewardship tasks identified by the NRC committee (NRC, 2003), will continue to press for a more "conservative" cleanup – cleanup that would be protective should be different, or more intensive land uses evolve in coming decades and centuries.

The chief finding of the National Research Council's Phase 2 Committee report captures the current Hanford situation very well:

> The committee has found no evidence that DOE (a) is considering requirements for and the likely effectiveness of LTS measures when establishing cleanup goals and approaches, or (b) has worked out practical and enduring means of implementing LTS so as to realize its goals for protection over the long term. In the recent emphasis by DOE on the objective of accelerating cleanup, the committee has seen no statement of how DOE will balance that objective against future risks. There is the possibility of a need for additional cleanup in the future if remediation is poorly planned or carried out. Explicit consideration of LTS issues when establishing cleanup goals and approaches would demonstrate that DOE is taking its responsibilities seriously – a key step in building trust among wary stakeholders. The failure to link LTS to cleanup undermines credibility and strengthens the fear among skeptical stakeholders and regulators that a hollow promise of stewardship is being imposed as a substitute for more costly and complete near-term cleanup (NRC, 2003).

NOTES

1. There are many accounts of the development of the Hanford site. Among the most readable summaries are M. S. Gerber, "Legend and Legacy: Fifty Years of

Defense Production at the Hanford Site" (Richland: Westinghouse Hanford Company, March 1992); M. S. Gerber, "The Hanford Site: An Anthology of Early Histories" (Richland: Westinghouse Hanford Company, October 1993); and Chapter 1 of Roy E. Gephart, "Hanford: A conversation about nuclear waste and cleanup" (Columbus and Richland: Battelle Press, 2003). See also F. G. Gosling, "The Manhattan Project: Making the Atomic Bomb" (Washington, DC: United States Department of Energy, January 1999).

2. The classic case of unreliability – often cited – is Love Canal, in Niagara Falls, NY. Closed in 1953 as a chemical landfill, it was sold to the Niagara Falls Board of Education and became an elementary school site, surrounded by homes, in spite of a recorded deed "warning" of the presence of chemical wastes. See "Background on the Love Canal," Love Canal Collection, State University of New York at Buffalo, Library Archives, available at http://ublib.buffalo.edu/libraries.

3. "The federal budget process allows an annual debate about national priorities that results in funding appropriations for long-term stewardship and all other government functions. A number of commentators have expressed the concern that the annual budget process does not provide guaranteed funding for long-term stewardship activities .... Because [other] alternatives ... would require specific Congressional action in the form of legislation or specialized appropriations, the viability of these alternatives is depending upon Congress concluding that the annual budget process is inadequate for this purpose" (DOE, 2001b).

4. Refer to draft "Long-Term Stewardship Site Transition Framework," transmitted to the State and Tribal Government Working Group by Assistant Secretary Jessie Hill Roberson on September 17, 2002.

5. See letter from A. Lynn Scarlett, Assistant Secretary of Interior, to Joseph Whitaker, Department of the Army, dated May 9, 2003. Assistant Secretary Scarlett makes it clear that it's not generally Interior's policy to accept land that has not been cleaned up, and then lays down several stringent requirements any granting agency must meet with regard to residual contamination (Scarlett, 2003).

6. "We believe there were a number of issues and key questions relevant to an assessment of institutional controls at the Hanford Site that were not addressed by the approach taken for this assessment." Letter from John B. Price, Washington Department of Ecology, and Mike Goldstein, EPA to Alex Teimouri, DOE Richland Operations Office, dated November 10, 2003. This appears to be a clear example of the situation in which " ... [T]he Department focuses narrowly on complying with regulations. While compliance is necessary, it is not sufficient, because today's regulations do not fully address LTS challenges" (NRC, 2003).

7. The author's observation is that the communities surrounding Hanford have always been acutely uncomfortable about dramatic depictions of its legacy of waste and contamination. A visit to Anaconda, Montana, found interpretive trails on old copper smelter areas, which tell of the history but make no mention of either the residual contamination or of the cleanup and remedies left in place to protect public health and the environment.

8. In a letter from Michael W. Owen, Director, Office of Legacy Management to Governors Gary Locke and Bill Owens, December 16, 2003: "Institutional controls can be a useful tool to ensure the protection of public health in sites that cannot be released for unrestricted use. However, we do not see them as critical to the

protection of public health at most DOE sites because we have emphasized the use of robust physical or engineered controls and sustained federal control of contaminated areas."

9. "The *Final Hanford Comprehensive Land-Use Plan Environment Impact Statement* … and the Presidential order creating the Hanford Reach National Monument which, together, identifies [sic] land uses following site cleanup provide a catalyst to re-evaluate the current cleanup baseline and Tri-Party Agreement Milestones to assure that the baseline will be in concert with the land use plans" (DOE, 2004a). One could argue to the contrary, however, that the stronger, more enduring the restrictions on uses in the Hanford Reach National Monument, the stronger will be pressures for population expansion and resource use on the balance of the Hanford site – the parts where significant contamination will remain.

10. At a June 24–35, 2004 workshop on 100-Area end states, the US Fish and Wildlife Service said that they expect to put resource managers on the land to be close to and learn about the resource. They also said that they expect to house fire fighters and maintenance workers within the monument. This is a typical practice elsewhere. These actions support continuing use of the residential farmer scenario for setting cleanup levels.

11. Skeptics will argue that the Department itself has existed for barely 30 years, that it has had nine secretaries, including one who, while in Congress, tried to abolish the Department, and that it has undergone frequent reorganizations, most of which have not greatly improved performance in the eyes of the General Accounting Office.

12. This decision concerning waste import will be re-examined in a new, comprehensive waste disposal Environmental Impact Statement under the terms of a settlement agreement reached by the State of Washington and the Department of Energy in January 2006. SETTLEMENT AGREEMENT re: *WASHINGTON v. BODMAN*, Civil No. 2:03-cv-05018-AAM January 6, 2006.

13. "DOE has yet to provide adequate assurances concerning closure and long-term stewardship of Hanford's disposal sites, including measures to monitor, maintain, and reassess remedies in light of new technologies or risks" (Hoffman, 2004).

REFERENCES

Bauer, C., & Probst, K. N. (2000). *Long term stewardship at contaminated sites: Trust funds as mechanisms for financing and oversight*. Washington, DC: Resources for the Future.

Brown, E. (2002). *Funding long-term stewardship of USDOE weapons sites: Tennessee's perpetual care trust fund*. Washington, DC: National Governors' Association Center for Best Practices.

Environmental Law Institute and Energy Communities Alliance. (2001). *The role of local governments in long-term stewardship at USDOE facilities*, Washington, DC.

Federal Register. (2000). Establishment of the Hanford Reach National Monument, Presidential Proclamation 7319 of June 9, 2000. Tuesday, June 13, 2000, pp. 37253–37256.

Geiser, D. (2004). Personal communication. Employee of DOE Office of Legacy Management, statement at a May 25, 2004 meeting of the National Governors Association Federal Facility Task Force held in Washington, DC.

Hanford Future Site Uses Working Group (Hanford FSUWG). (1992). *The future for Hanford: Uses and cleanup*. Richland, WA, December.

Harris, S. (1998). Cultural legacies. Paper delivered at society for risk analysis annual meeting in Phoenix, AZ, December 7, 1998.

Hoffman, L. (Director of Washington Department of Ecology). (2004). Letter to Jessie Hill Roberson (Assistant Secretary of Energy for Environmental Management) March 9, 2004.

Locke, G. G., & Owens, G. B. (2003). Letter to Michael Owen, November 12.

Miller, D. S. (2002). Looking a gift horse in the mouth: Federal Agency Opposition to State Institutional Control Laws. *Environmental Law Reporter*, *32*(9), 11115–11126.

National Research Council (NRC). (2000). *Long-term institutional management of U.S. Department of Energy legacy waste sites*. Washington, DC: National Academy Press.

National Research Council. (2003). *Long-term stewardship of DOE legacy waste sites – a status report*. Washington, DC: National Academy Press.

Owen, M. W. (Director, Office of Legacy Management). (2003). Letter to Governors Gary Locke and Bill Owens, December 16, 2003.

Owendoff, J. (1999). (DOE Acting Assistant Secretary for Environmental Management). Letter to co-Conveners of the State and Tribal Government Working Group, May 24, 1999.

Price, J. B. (Washington Department of Ecology) and Mike Goldstein (USEPA). (2003). Letter to Alex Teimouri (USDOE Richland Operations Office), November 10, 2003.

Price, J. B., & Power, M. S. (2002). A Washington State Perspective on long-term stewardship at Hanford. Paper delivered at waste management '02 in Tucson, AZ.

Scarlett, A. L. (2003). Letter from A. Lynn Scarlett, Assistant Secretary of Interior, to Joseph Whitaker, Department of the Army, May 9, 2003.

State & Tribal Government Working Group (STGWG). (2001). *Long-term stewardship committee*. Interim report on information management for long-term stewardship, October.

US Department of Energy. (1991). *Decommissioning of eight surplus production reactors at the Hanford site*. Richland, WA: Department of Energy. DOE/EIS 0119.

US Department of Energy (DOE). (1992). *The future for Hanford: Uses and cleanup*. Richland, WA: Department of Energy.

US Department of Energy. (1996). *Plutonium: The first 50 years*. Washington, DC: United States Department of Energy.

US Department of Energy. (1997). *Linking legacies: Connecting the cold war nuclear weapons production processes to their environmental consequences*. Washington, DC: United States Department of Energy–Office of Environmental Management.

US Department of Energy. (1999). *Final Hanford comprehensive land-use plan environmental impact statement*. Richland, WA: Department of Energy. DOE/EIS-0222-F.

US Department of Energy. (2001a). *Report to Congress: Long-term stewardship report*. Washington, DC: U.S. Department of Energy.

US Department of Energy. (2001b). *Long-term stewardship study: Volume 1 – report*. Washington, DC: U.S. Department of Energy, Office of Environmental Management.

US Department of Energy. (2002a). *Performance management plan for the accelerated cleanup of the Hanford site*. Richland, WA: United States Department of Energy-Richland Operations Office.

US Department of Energy. (2002b). *A review of the environmental management program*. Washington, DC: Department of Energy. February.

US Department of Energy. (2003a). *Hanford federal facility agreement and consent order, as amended through April 18, 2003*. Richland, WA: United States Department of Energy – Richland Operations Office. Document 89-10 Rev. 6, April.

US Department of Energy. (2003b). *Hanford long-term stewardship program and transition: Preparing for environmental management cleanup completion*. Richland, WA: Department of Energy August.

US Department of Energy. (2003c). Office of Legacy Mgmt Strategic Plan.

US Department of Energy. (2004a). *Hanford site risk-based end state vision (draft)*. Richland, WA: Department of Energy. DOE/RL-2003-59. April. Available at http://www.hanford.gov/docs/rbes/PDF/RBES.pdf

US Department of Energy. (2004b). *Record of decision for the solid waste program, Hanford site, Richland, Washington*. Washington, DC: U.S. Department of Energy June.

US Department of Energy (2005). *Hanford Site Risk-Based End State*. Richland, WA: Department of Energy. DOE/RL-2005-57. October, 2005, Available at http://www.hanford.gov/docs/rbes/final.cfm

US EPA. (1995). *Declaration of the record of decision, USDOE Hanford 100 area, 100-BC-1 100-DR-1 and 100-HR-1 operable units, Hanford site, Benton County, Washington*. Richland, WA: U.S. Environmental Protection Agency.

Weapons Complex Monitor. (2004). Waste reclassification language remains in Senate defense bill. Vol. 15, #23, June 7, 2004, pp. 2–4.

CERCLA REMEDY SELECTION LESSONS: THE SHATTUCK SITE

Elizabeth K. Hocking

INTRODUCTION

Remedies for the nation's contaminated lands must be protective of human health and the environment through time in order to comply with the Comprehensive Environmental Response, Compensation, and Liability Act (CERCLA) and other environmental laws. The high cost associated with remediating these lands – between $8 billion and $35 billion to address 15 million acres possibly contaminated with military munitions (U.S. General Accounting Office, 2003), more than $7 billion in 2004 alone for environmental restoration at U.S. Department of Energy sites (DOE, 2004), and uncalculated costs for the private sector – encourages decision-makers to select cost-effective remedies.

The remedy selection odyssey at the Shattuck Site, in which the remedy selected in 1992 and constructed in 1997 was completely rejected in 2000, presents several cautionary lessons on selecting remedies: the area of concern addressed by the remedy should take into account past, present, and reasonable future resource use of the contaminated area and contiguous areas; reliance on institutional controls to minimize or restrict resource use may be misplaced; and community acceptance of the remedy is of the utmost importance. Although the Shattuck Site is not a Federally owned site, the lessons learned apply to both private parties and Federal decision-makers

Long-Term Management of Contaminated Sites
Research in Social Problems and Public Policy, Volume 13, 105–114

ISSN: 0196-1152/doi:10.1016/S0196-1152(06)13004-5

and are a reminder that efforts to minimize the expense of remediation that neglect remedy protectiveness can prove short-sighted and costly.

SHATTUCK SITE BACKGROUND

Shattuck Site History

The Shattuck Site, also known as Operable Unit VIII (OU VIII) of the Denver Radium Site, is a 10-acre site within the City of Denver. It is in an area of the city designated for commercial/industrial use; however, three blocks from the site are residential areas.

The Shattuck Site was used for several mineral-processing operations for approximately 60 years. Site operations included treatment of molybdenum ores, extraction of ferric vanadate, processing of radium slimes for radium recovery, and processing of uranium compounds, molybdenum, uranium ores, and rhenium.

Shattuck Remediation Decision History

The Shattuck Site was placed on the Comprehensive Environmental Response, Compensation, and Liability Act (CERCLA) National Priorities List (NPL) in 1983. Soil at the site was considered contaminated if the radium-226 concentration exceeded 5 pCi/g above background in the top 15 cm of the soil, or 15 pCi/g above background in any layer below the top 15 cm. The volume of contaminated soil was estimated at 50,000 yd^3. The principal health threat posed by the site was the carcinogenic risk related to the radium-226 in the soil. Thorium-230, uranium, selenium, and other heavy metals were also contaminants of concern because they exceeded health-based action levels, but were primarily co-located with the radium-226 soils.

In March 1991, the U.S. Environmental Protection Agency (EPA) and the Colorado Department of Health proposed a remediation plan for the site that described remedial action alternatives: no action; restricted site access with excavation of some soil; capping with clean soil; in situ vitrification; excavation, stabilization, and on-site disposal; excavation, removal, and replacement of contaminated soil; and excavation and physical or chemical soil treatment. The proposed plan identified excavation, removal, and replacement as the preferred remedy for the Shattuck Site and requested public comment on it as well as on the other remedial action alternatives.

In the January 1992 CERCLA Record of Decision (ROD) for the site, the EPA rejected its earlier preferred remedial alternative and selected the alternative of excavation, stabilization, and on-site disposal of the contaminated soils, along with capping and institutional controls.[1] The institutional controls element of the remedy was the development of a program to control access to the stabilized materials through deed restrictions or annotations denoting that (1) the property is dedicated as a waste disposal site and that (2) excavating into the cover and stabilized materials, constructing enclosed structures on the disposal site, using groundwater located beneath the site, and using the site for agricultural purposes are restricted. The property owner was also told to "use reasonable efforts to seek zoning restrictions from the City and County of Denver as further assurance that the institutional requirements associated with the site will not inadvertently be breached in the future" (EPA, 1992).

In explaining why the selected remedial alternative was different from the preferred alternative, the EPA stated that the "most decisive factors" in selecting stabilization were that it better satisfied the CERCLA preference for the "reduction of toxicity, mobility, or volume through treatment"[2] and the cost-balancing criteria of the National Contingency Plan (NCP) (EPA, 1992). Accordingly, the EPA issued a remedial order requiring the Shattuck owners to perform the selected remedy.

The City of Denver was displeased with the selected remedy and issued a cease and desist order to the owner of the site in May 1994, claiming that the owner's EPA-mandated action would violate City zoning ordinances that prohibit maintenance of hazardous waste in areas zoned for industrial use. Ensuing discussions between the City and the EPA to resolve the City's concern were unsuccessful and resulted in the EPA suing the City in August of that year. In February 1996, the U.S. District Court enjoined the City from enforcing its zoning ordinance against the Shattuck owners; the City lost its appeal of that decision in November 1996 (US v. City and County of Denver, 1996, 44 ERC 1121). The excavation and stabilization of contaminated soil at the Shattuck Site was completed in November 1997.

The result of the stabilization was a monolith that stood 12–15 ft above street level. The volume of the monolith was 83,610 yd^3, 43,214 of which were the contaminated soils excavated from the site, with the remainder being fly ash and cement (EPA, 2000a).

The community supported the EPA's proposed preferred remedial alternative, excavation and removal, and opposed the selected remedy. Several community groups sustained their opposition from 1992 onwards and worked actively to have the remedy overturned (Raimond, 2001; EPA, 1999).

Five-Year Review of Shattuck Site Remedy

Five-year remedy reviews are required by CERCLA when selected remedial actions result in "any hazardous substances, pollutants, or contaminants remaining at the site" and are conducted "to assure that human health and the environment are being protected" (CERCLA, 1980). The EPA guidance on conducting five-year reviews describes their purpose as evaluating remedy implementation and performance to determine if the remedy "is or will be protective of human health and the environment" (EPA, 2001).

The first five-year CERCLA review of the Shattuck Site after the remedy was fully implemented was conducted in 1999. That review identified deficiencies in the monolith-monitoring plan, plume-monitoring plan, site characterization and modeling, and the risk assessment that formed the basis for developing the remedial alternatives. Additional deficiencies were the lack of institutional control of the plume outside the site boundary, vulnerability of the cover and monolith design to long-term degradation, and several specific technical design issues (EPA, 1999). Each deficiency was judged as currently affecting remedy protectiveness.

ROD Amendment for the Shattuck Site

In addition to the remedy effectiveness concerns raised in the 1999 five-year review, the local community as well as the city and county requested that the EPA consider remedy alternatives that would allow unrestricted use of the Shattuck Site. The State of Colorado, which had agreed in 1992 to the on-site stabilization remedy, joined in the request.

In 2000, the EPA issued a ROD Amendment indicating that the new selected remedy for the Shattuck Site was the removal of the monolith and excavation of any identified contaminants. The authors of the amended ROD stated that the CERCLA preference for treatment, cited in the 1992 ROD as one of the decisive factors for selecting on-site stabilization, had already been satisfied because the contaminated soil had been treated as part of the stabilization process to form the monolith. The other decisive factor that influenced the 1992 remedy selection was cost. The preferred remedial alternative was rejected because it was not as cost-effective as the stabilization remedy. In the ROD Amendment, the EPA stated that even though off-site removal of the stabilized material was the most costly alternative, it was still considered the most cost-effective because of the certainty it provided in the protection of human health and the environment.

In choosing removal of the stabilized material as the new selected remedy, the EPA described it as being the alternative that is "least dependent upon ICs [institutional controls] that historically have been difficult to implement" (EPA, 2000a). Removing the monolith will allow unrestricted development and use of the land.

REMEDY SELECTION LESSONS FROM THE SHATTUCK SITE

Three of the remedy deficiencies identified in the five-year review – site characterization, risk assessment, and institutional control plan – are linked and originated in the lack of consideration given to past, present, and reasonable future land and water use of the area surrounding the Shattuck Site when the original remedy was selected. The site characterization was inappropriately limited to the Shattuck Site boundary, which led to the risk assessment that failed to take into account exposure pathways and reasonable land uses, resulting in development of an institutional control program that did not address off-site land or water use. Consequently, the remedy was quickly judged to have failed to provide overall protection of human health and the environment. These deficiencies would have been avoided if public opinion had been given greater consideration when the original remedy was chosen.

Site Characterization

One way to describe the problem with the original site characterization is that it looked only at the Shattuck Site rather than the Shattuck Site in conjunction with the outlying area. In this context, the Shattuck Site is the 10-acre geographic area listed on the NPL as OU VIII of the Denver Radium Site. The "Shattuck site" is the larger geographic area surrounding OU VIII. The Shattuck Site is the pebble dropped into the pond; the "Shattuck site" the ripples from the pebble. The pebble was characterized but its rippling effect was not.

The "Shattuck site" is within the City of Denver and includes industrial, commercial, recreational, and residential uses. Surface water is used for recreation. There are groundwater users in the "Shattuck Site" although groundwater is not known to be used for drinking. The aquifer under the Shattuck Site is considered a potential source of drinking water, however, and there is a history of groundwater use in the area. While there were no known downgradient groundwater users, the Shattuck Site is located over a shallow, unconfined aquifer that merges with the alluvial aquifer beneath the South Platte River floodplain.

Risk Assessment

The five-year reviewers questioned the adequacy of the original risk assessment on two fronts. First, despite the evidence of some water use and the presence of the aquifer, the reviewers stated that the risk assessment did not include any groundwater and surface-water exposure scenarios (EPA, 1999). Second, even though the land use around the Shattuck Site was mixed, the industrial land use standard was the only basis for establishing remediation goals. The use of that standard was possibly shortsighted given that the Shattuck Site and surrounding land were located in the City of Denver, an area that could reasonably be expected to experience population growth and economic development involving varied uses.

Institutional Controls

The institutional controls element of the remedy – control of access to the stabilized material and on-site groundwater – flowed naturally from the Shattuck Site focus of the risk assessment. Accordingly, the Shattuck Site landowner was responsible for developing the deed restrictions that would limit site and groundwater use and for making reasonable efforts to request zoning restrictions from the city as a further safeguard to restrict resource use.

These institutional controls were perfectly reasonable for the Shattuck Site. They were, however, unreasonable for the "Shattuck site" – the outlying area – because they provided no control over potential access to contaminated off-site water.

The only way to limit access to any off-site contaminated water was for the local government to use its authority to enact an ordinance controlling the use of groundwater downgradient from the Shattuck Site. Authors of the five-year review observed that it was "questionable" whether the institutional controls necessary to restrict the use of off-site groundwater could ever be implemented. The EPA removed any doubt about that point by stating in the ROD Amendment that the City and County of Denver "has indicated that it will not enact such an ordinance" (EPA, 2000a). The EPA concluded that the nonimplementability of an institutional control necessary to ensure remedy protectiveness helped justify removal and off-site disposal of the stabilized contaminated soil.

Community Acceptance

In the summary of responses to the significant comments made during the comment period on the original preferred remedial alternative (i.e., excavation and off-site disposal), the EPA recognized that "the City and County of Denver and the nearby community strongly support excavation and off-site disposal and strongly object to alternatives that would leave the contaminated material on-site" (EPA, 1992).

This early on-the-record objection by the local government and the City's subsequent lawsuit challenging the implementation of the selected remedy go far in explaining why the institutional control ordinance would not be enacted. To do so would restrict future development of the area and abet a remedy that the community did not support.

Community members did indeed resist the remedy selected in 1992. They applied for and received an EPA Technology Assistance Grant to enable them to review the effectiveness of the remedy, filed a case in the U.S. District Court challenging the stabilization remedy, and requested that the EPA review that selected remedy.

The five-year review authors found that the public's goals were not adequately "assessed and reflected" in the selection of the original remedy in 1992 and suggested that the remediation concept be re-evaluated to take into account the community's land use goals (EPA, 1999). The ROD Amendment authors concurred with that finding, stating that the EPA's most recent interactions with the public "cast doubt on the certainty in the original assumptions made regarding future land and groundwater use at the Site" (EPA, 2000a).

SELECTING CERCLA REMEDIES

The planning that supported the risk assessment, remedial goals, and remedial alternatives developed for the 1992 ROD simply did not go far enough temporally or geographically. For paying insufficient attention to the actual and potential future uses of the land and water surrounding the Shattuck Site, the selected remedy was categorized as being inadequately protective within eight years of its selection.

Remediation decision-makers consistently face the challenges of decreasing the cost of cleanup but increasing the speed of cleanup. The resulting pressure from these challenges can lead to defining the geographic area of remediation concern too narrowly and ultimately failing to meet CERCLA's requirement that remedies protect human health and the environment. This tendency could be avoided by assessing the impacts of remedial alternatives on present and reasonably expected future resource use of both the contaminated area and the surrounding affected areas.

That will require adequate characterization of the contaminated site and contiguous areas in terms of past, present, and reasonable future land and water uses. Armed with this information, risk assessors are in a better position to identify and evaluate all reasonable exposure pathways, which should lead to the development of realistic remediation goals and remedial alternatives.

The actual remedy selection process will still be guided (as it was when the 1992 Shattuck Site remedy was chosen) by the criteria listed in the NCP. These are divided into three categories (NCP, 1968):

- Threshold criteria
  - Overall protection of human health and the environment, and
  - Compliance with Applicable Relevant and Appropriate Requirements (ARARs).
- Primary balancing criteria
  - Long-term effectiveness;
  - Reduction of toxicity, mobility, and volume through treatment;
  - Short-term effectiveness;
  - Implementability of the remedy; and
  - Cost of the remedy.
- Modifying criteria
  - State acceptance of the selected alternative, and
  - Community acceptance of the selected alternative.

The implementability criterion is especially important when evaluating remedial alternatives that include leaving hazardous substances, pollutants, or contaminants on site. Unless the Federal government retains the land, these remedial alternatives will almost always require some type of institutional control. If the institutional control cannot or, as in the Shattuck case, will not be put into effect, the alternative will have to be eliminated because it fails the NCP's implementability criterion. Although community acceptance of the remedy is listed as the last of the nine remedy selection criteria, it is certainly not the least important. The concerted, on-going, and ultimately successful efforts of the local government and community members to resist and reject the original Shattuck remedy demonstrate the importance of community acceptance.

Cost of the remedy is another selection criterion. The up-front effort to identify reasonable remedial alternatives through thorough characterization of the contaminated area and surrounding areas may require more time and money, and it may result in more costly remedial alternatives. But scrimping on the effort by too narrowly focusing the area of concern can take its own toll.

In addition to being overturned in the space of eight years and causing substantial social cost in the form of lack of trust in the EPA (Raimond, 2001), the Shattuck Site remedy eventually cost approximately $48.1 million: $26.6 million paid by the owner to stabilize the contaminated material on-site, plus $21.5 million to remove the stabilized soil and dispose of it off-site.[3] The 1991 preferred remedial alternative – excavation and disposal off-site – was rejected in part because it was less cost-effective "in proportion to its overall effectiveness" (EPA, 1992). The estimated cost of that preferred alternative was $36.8–$48.8 million.

NOTES

1. The EPA defines institutional controls as "non-engineered instruments such as administrative and/or legal controls that minimize the potential for human exposure to contamination by limiting land or resource use" (EPA, 2000b).

2. "Remedial actions in which treatment which permanently and significantly reduces the volume, toxicity or mobility of the hazardous substances, pollutants, and contaminants is a principal element, are to be preferred over remedial actions not involving such treatment. The offsite transport and disposal of hazardous substances or contaminated materials without such treatment should be the least favored alternative remedial action where practicable treatment technologies are available" (CERCLA, 1980).

3. The EPA assumed 90% of the cost and the State of Colorado will pay the remaining 10% out of funds received from the Shattuck Chemical Company (Colorado Department of Public Health and Environment, 2003).

ACKNOWLEDGMENTS

Work supported by the U.S. Department of Energy under contract W-31-109-Eng-38.

The submitted manuscript has been created by the University of Chicago as Operator of Argonne National Laboratory ("Argonne") under contract

No. W-31-109-ENG-38 with the U.S. Department of Energy. The U.S. Government retains for itself, and others acting on its behalf, a paid-up, nonexclusive, irrevocable worldwide license in said article to reproduce, prepare derivative works, distribute copies to the public, and perform publicly and display publicly, by or on behalf of the Government.

REFERENCES

Colorado Department of Public Health and Environment. (2003). *Shattuck superfund site: A fact sheet.* July.

CERCLA (Comprehensive Environmental Response, Compensation, and Liability Act). (1980). Public Law 96-150, as amended, 94 Stat. 2767, 42 USC 9601 et seq.

DOE. (2004). *FY 2005 Congressional Budget Request*, DOE/ME-0036, Vol. 5, February.

Environmental Protection Agency (EPA). (1992). *Superfund Record of Decision: Denver Radium Site*, EPA ID: COD980716955, OU8, Denver, CO, EPA/ROD/R08-92/063, January.

EPA. (1999). *Five-year review report.* SC&A, Inc., Denver, CO, November.

EPA. (2000a). *Superfund Record of Decision Amendment: Denver Radium Site*, EPA ID: COD980716955, OU8, Denver, CO, EPA/AMD/R08-00/561, June.

EPA. (2000b). *Institutional controls: A site manager's guide to identifying, evaluating and selecting institutional controls at Superfund and RCRA Corrective Action Cleanups*, EPA 540-F-00-005, OSWER 9355.0-74FS-P, September.

EPA. (2001). *Comprehensive five-year review guidance*, EPA 540-R-01-007, June.

National Oil and Hazardous Substances Pollution Contingency Plan (NCP). (1968). 40 CFR Part 300, Section 430, Remedial Investigation/Feasibility Study and Selection of Remedy, *Code of Federal Regulations.*

Raimond, R. R. (2001). *Trust and conflict in public participation.* Division of Hazardous Materials and Waste Management, Colorado Department of Public Health and Environment, January.

United States v. City and County of Denver. (1996). 44 ERC 1121 (10th Cir. 1996).

U.S. General Accounting Office. (2003). *Military munitions: DOD needs to develop a comprehensive approach for cleaning up contaminated sites*, GAO-04-147, December.

SCIENTIFIC AND TECHNICAL PERSPECTIVES ON LONG-TERM MANAGEMENT

A SUSTAINABLE ENVIRONMENTAL PROTECTION SYSTEM FOR THE MANAGEMENT OF RESIDUAL CONTAMINANTS

Kevin M. Kostelnik, James H. Clarke, Jerry L. Harbour, Florence Sanchez and Frank L. Parker

ABSTRACT

Hazardous and radioactive wastes are routinely disposed of in shallow land disposal facilities at sites that have undergone environmental remediation. These residual contaminants, which remain on-site in these isolation facilities, continue to pose risks to humans and the environment and represent extremely long-term liabilities that require continuous management.

This paper draws upon the experiences generated from two federal programs established in response to the Uranium Mill Tailings Radiation Control Act (UMTRCA) of 1978 and the Comprehensive Environmental Response, Compensation, and Liability Act (CERCLA) of 1980. Operational issues arising at existing sites suggest that there is a need to better integrate the management functions at contaminant isolation facilities. To protect human health and safeguard the natural environment, more sustainable environmental protection systems are required for the

Long-Term Management of Contaminated Sites
Research in Social Problems and Public Policy, Volume 13, 117–137

ISSN: 0196-1152/doi:10.1016/S0196-1152(06)13005-7

long-term management of residual contaminants. A series of logic diagrams are introduced to improve the integration and management of remedial processes, engineered barriers and institutional controls. These institutional responsibilities need to be monitored and maintained by the appropriate site stewards to ensure continued system performance.

INTRODUCTION

Scientific, engineering, and social advancements have significantly raised the standard of living for most of the world's population. These advancements, however, have also produced adverse impacts. One such impact is the environmental degradation associated with the necessary disposal of certain by-products, such as hazardous and radioactive waste materials.

Hazardous, radioactive, and other toxic substances are the unwanted result of energy production, mineral extraction, national defense programs, as well as industrial and manufacturing operations. In many circumstances, processes do not exist that can destroy these materials; thus they are routinely disposed of in shallow burial grounds throughout the world. This disposal practice has restricted and limited the use of private and public properties, has increased the risk to human health and the environment, and has resulted in damage to ecosystems by contaminating considerable volumes of soil and groundwater. The magnitude of such environmental problems is enormous and affects international, national, state, and local entities.

The United States is addressing these problems by enforcing a variety of environmental regulations. Two primary regulations applicable to the remediation of abandoned radioactive and hazardous waste sites are the Uranium Mill Tailings Radiation Control Act (UMTRCA) of 1978 and the Comprehensive Environmental Response, Compensation, and Liability Act (CERCLA) of 1980. UMTRCA defines the remediation requirements for abandoned sites that had processed uranium and related ores for the federal government (UMTRCA, 1978). CERCLA, commonly known as Superfund, establishes the framework for the federal response to the release of hazardous substances that endanger public health or the environment (CERCLA, 1980).

Additional related regulations include the Resources Conservation and Recovery Act (RCRA) of 1976 (RCRA, 1976) and the Small Business Liability Relief and Brownfields Revitalization Act (Brownfields) of 2002 (USEPA, 2002). RCRA establishes requirements for managing hazardous, industrial, and household wastes from generation through disposal to minimize future pollution that may result from solid waste landfills. The

Brownfields Program, signed into law January 11, 2002 (USEPA, 2002) focuses on abandoned, idle, or under-utilized industrial and commercial facilities where expansion or redevelopment is complicated by real or perceived environmental contamination.

Despite these regulatory advancements, environmental problems persist. Society is facing the realization that it cannot restore many of the world's environmentally contaminated sites to pristine conditions (NRC, 2000) because many of today's waste management techniques do not eliminate the waste contaminants. Rather, they only concentrate or attempt to contain the hazardous constituents (NRC, 1997; Russell, 2000).

To protect human health and safeguard the natural environment, a sustainable contaminant isolation system is required for the long-term management of residual contaminants (NRC, 2000; INEEL, 2003). Residual contaminants result from the presence of hazardous and/or contaminated material that remains on-site following active operations or the completion of remedial actions. These residual contaminants pose continued risk to humans and the environment and represent extremely long-term liabilities (i.e., 100's–1000's of years) that require continuous management.

CONTAMINANT ISOLATION CONCEPTUAL MODEL

Management strategies for residual contaminants typically involve the emplacement of the contaminated materials into near-surface isolation facilities and monitoring of the facility and surrounding groundwater for the long-term. The best available technology of the time is most often used in the design, construction, and monitoring of these facilities. The intent of these facilities is to maintain the long-term isolation of the identified contaminants (i.e., radioactive, organic, inorganic, etc.) as well as to mitigate their associated hazards (i.e., gamma radiation, radon emanation, contaminant migration, fire, and explosion potential, etc.).

A typical near-surface contaminant isolation facility consists of several engineered components as well as institutional controls. The engineered components include a surface cover, the stabilized hazardous material, a monitoring system and in some cases subsurface barriers (i.e., bottom liners and side walls), and a leachate recovery system (Benson et al., 2002; Clarke et al., 2004). Institutional controls, such as government controls (i.e., zoning and groundwater use restrictions) and property-based controls (i.e., deed restrictions, covenants), are often incorporated into the management plan for sites with residual contaminants (Borinsky, 1995; Gaspar & Burik, 1998).

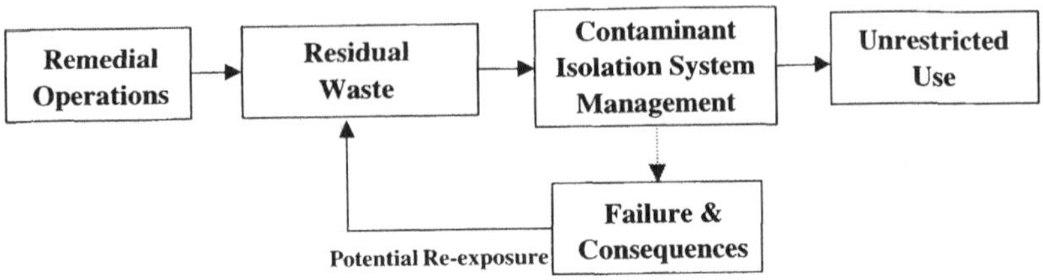

Fig. 1. Conceptual Model for Residual Contaminant Isolation Management.

Fig. 1 illustrates the process often used for the management of residual contaminants. Remedial operations consolidate contaminated materials on-site and contain these residual contaminants within an isolation facility. The objective of this facility, and its associated management, is to isolate these contaminants for hundreds to thousands of years or until the contaminants no longer present unacceptable risk to humans or the environment. The site could potentially be released for unrestricted use at that time if the contaminants decay and the associated site risk is reduced to acceptable levels.

Society's experience with modern waste isolation techniques, however, is beginning to show that waste isolation facilities and management techniques do not always perform as expected (USGAO, 1990). The reality is that facility performance can significantly deviate from planned performance. If not rectified by the site stewards, these performance deviations can negatively affect facility performance to the point that the system fails, resulting in negative consequences such as receptors again being exposed to the residual contaminants.

Fig. 1 illustrates the significance of the contaminant isolation facility management step in isolating contaminants. Several important questions can be asked of this step. Namely, why and how do management efforts deviate from original plans? When do these deviations lead to contaminant isolation facility failure versus continued isolation? Are there precursors to failure? This paper illustrates how an analytical approach can help provide answers to these critical questions.

The Case of Love Canal

Love Canal serves as a classic case study with regard to the failure of isolating residual contaminants for the long term. Located in Niagara Falls, New York, the site was initially developed by William T. Love to serve as a canal between the upper and lower reaches of the Niagara Falls. Although this vision was never realized the site's low-permeability clay soils later made the site appealing for containing hazardous waste.

Hooker Electrochemical Co. [(Hooker) now Occidental Chemical Corporation] and others disposed of waste at the site from 1942 until 1953 (Mazur, 1998). The waste disposed of by Hooker consisted primarily of liquid and solid chemical residues and by-products from Hooker's chemical manufacturing processes. The U.S. Environmental Protection Agency (USEPA) concluded that approximately 21,000 tons of chemicals were disposed of in the canal (USEPA, 1991).

Environmental and public health concerns began to surface in the mid to late 1970s (Brown, 1980; Zuesse, 1981; Mazur, 1998) following several years of above average precipitation. In August of 1978 a health emergency was declared and portions of the population surrounding Love Canal were evacuated. In May of 1980, President Carter declared the site a national disaster, resulting in a second evacuation.

Several important developments and key activities appear to have contributed to the facility failure and the subsequent health emergencies. The property containing the waste was transferred from private to public ownership. Property-based deed notices were included in the initial land transfer (NYS, 1953). Demographic patterns indicate population migration toward the disposal site. Portions of the site property were subsequently transferred. Infrastructure construction modified the contaminant isolation cell. And, information concerning the previous use of the site was not readily available or maintained with local stakeholders.

Additional information on Love Canal is presented in many books, articles and reports (Zuesse, 1981; Gibbs & Levine, 1982; Whelan, 1985; NYS, 1988; ETF, 1998; Gibbs, 1998; Thompson, 2004). Additionally, the USEPA provides official details of the site's remediation (USEPA 1988, 1991, 2003a, 2003b).

SYSTEM LOGIC

To reduce the likelihood of failure, the management of residual contaminants requires multiple and redundant controls. Engineered barriers are controls and practices that primarily modify the natural setting and the ecosystem. Institutional controls are practices designed to modify human behavior. Combinations of both types of controls are believed to improve the likelihood that residual contaminants will remain isolated from potential receptors. This concept of layering and defense-in-depth is advocated in numerous studies (USDOD, 1998; NRC, 2000; USEPA, 2000; INEEL, 2003).

As a means of investigating the performance of engineered barriers, Sanchez, Clarke, and Parker (2002) initiated the development of a series of logic diagrams. This work is now being expanded to investigate the total contaminant isolation system. The following logic diagrams define the actions that are continually required for the management step to remain consistent with its plan and thus avoid deviations that may lead to system failure. These diagrams provide an adaptive means for site stewards to reassess the situation and make

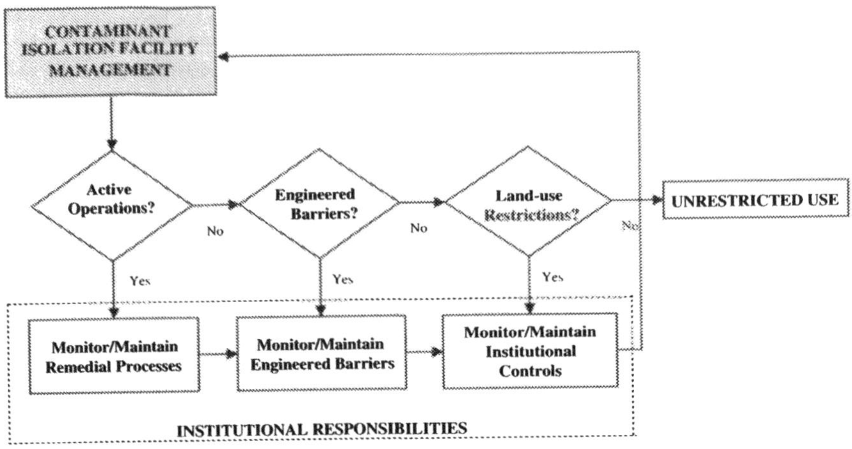

Fig. 2. Level 1 Logic Diagram for the Management of a Contaminant Isolation Facility.

necessary adjustments to account for spatial and temporal changes, which potentially could impact the system.

Fig. 2 illustrates the top-level (Level 1) logic diagram for the contaminant isolation facility management step. This level involves asking three primary questions. Does the facility have:

(a) active operations,
(b) engineered barriers, or
(c) land-use restrictions?

As the operation of the contaminant isolation facility may change over time, these questions need to be routinely asked as site stewards continually cycle through the logic diagram. An affirmative response to any of these questions triggers site-specific institutional requirements that must be fulfilled to ensure the continued performance of the contaminant isolation facility.

Institutional Responsibilities

Institutional responsibilities can be viewed as a series of critical functions and activities that must be continually conducted to ensure that the remedial system performs as designed. The functions of a sustainable contaminant isolation system, as illustrated in Fig. 2, include:

- maintaining active remedial processes,
- maintaining the engineered barriers, and
- maintaining institutional controls.

Remedial Processes

Remedial systems often include an active remediation component. These active remedial processes are commonly associated with the ongoing treatment of contaminated groundwater or a treatable source-term in the vadose zone. Representative examples include pumping and ex-situ treatment of contaminated groundwater, in situ bio-remediation of contaminated groundwater, natural attenuation (i.e., passive remediation), and vapor vacuum extraction of volatile organic compounds (Looney & Falta, 2000; Suthersan, 2002).

These processes require continued operational support to ensure they perform as expected. Power supplies, equipment maintenance and change-out, operating procedures and protocols, groundwater sampling, sample management, and analysis are a few of the common support activities required if active remedial processes are implemented as a part of a selected remedy.

While these techniques have been shown to be effective given favorable site conditions, considerable effort is required for their implementation (NRC, 1997; USEPA, 1998). For example, significant volumes of groundwater may need to be pumped and treated. In addition, these operations may be required for decades to achieve the anticipated risk reduction. Fig. 3 illustrates the level 2 Logic Diagram for the Remedial Processes component of the contaminant isolation facility management step.

Engineered Barriers

Engineered barriers are structures designed to modify the natural setting. Near-surface contaminant isolation facilities typically consist of several engineered barriers: a surface cover, the stabilized contaminated material, a monitoring system, and in some cases a leachate recovery system and subsurface barriers (i.e., bottom liners and side walls). These engineered barriers require periodic maintenance to ensure they remain effective.

The surface cover (i.e., surface liner) of a near-surface contaminant isolation facility serves as the primary barrier and is intended to perform two critical functions:

(A) The prevention of precipitation or surface water infiltration into the contaminated materials, and
(B) The prevention of direct contact with these constituents.

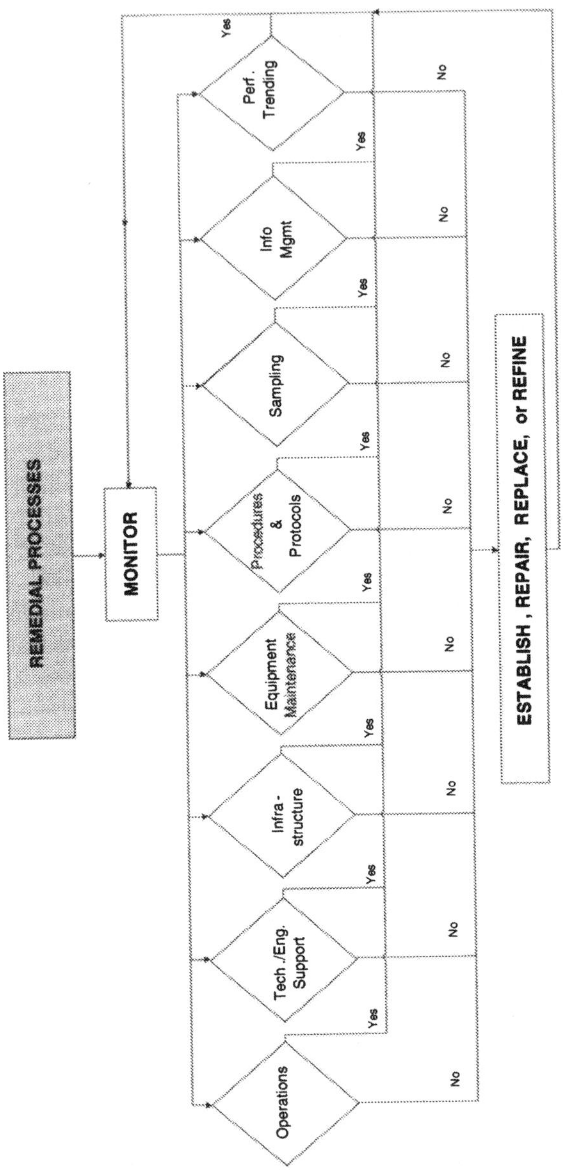

Fig. 3. Level 2 Logic Diagram for the Remedial Processes Component of the Contaminant Isolation Facility Management Step.

These functions are achieved through the use of specific barrier layers, including erosion protection barriers, surface water drainage layers, freeze–thaw protection layers, and structural stability layers. Also, in some cases such as remediated uranium mill tailing sites, the cover serves as protective barrier for radon gas.

Designs utilize both natural and synthetic materials via a multiple-layer approach. Maintenance efforts for surface covers include repairs to damaged layers resulting from surface erosion or bio-intrusion, re-contouring surface features, vegetative controls (i.e., cutting, spraying herbicide, etc.), and overall site monitoring.

In addition, engineered barriers include the hazardous material (e.g., tailings, waste, debris) placed in the contaminant isolation facility. Current regulations require this material to be in a stable form (i.e., compacted or otherwise stabilized) to reduce the potential for post-closure subsidence and settlement and to minimize porosity and control permeability within the facility. A leachate recovery system and subsurface barriers (i.e., bottom liners) if used, is predicated on site hydrology, ecology, climate, local groundwater characteristics and use patterns, and regulatory requirements. Maintenance of these components generally consists of monitoring their continued performance. Failure of these components may require repair, replacement or re-remediation of the site. Fig. 4 illustrates the Level 2 logic diagram for the continual assessment of engineered barrier maintenance efforts.

Institutional Controls

Institutional controls are processes, instruments, and mechanisms designed to influence human behavior and activity. These mechanisms generally take the form of legal or administrative restrictions. Institutional controls can be classified into three primary mechanisms: government controls, proprietary controls, and public notice (DERTF, 1996; Breggin, Pendergrass, & McElfish, 1998; Pendergrass, 1999). Government controls include zoning, groundwater restrictions, siting restrictions, and ordinances (Clancy-Hepburn et al., 1995; Edwards, 1997, 2000a). Proprietary controls are common-law property-based restrictions and include deed restrictions, covenants, reversionary interests, and easements (Pendergrass, 1999). Public notice involves the dissemination of information and includes broad public publication, state registries, and signage (DERTF, 1996; Breggin et al. 1998; Pendergrass 1999).

In addition to these conventional institutional controls, the site steward (i.e., responsible party/institution) must maintain the security of the site from inadvertent or intentional intrusion, maintain financial security of the

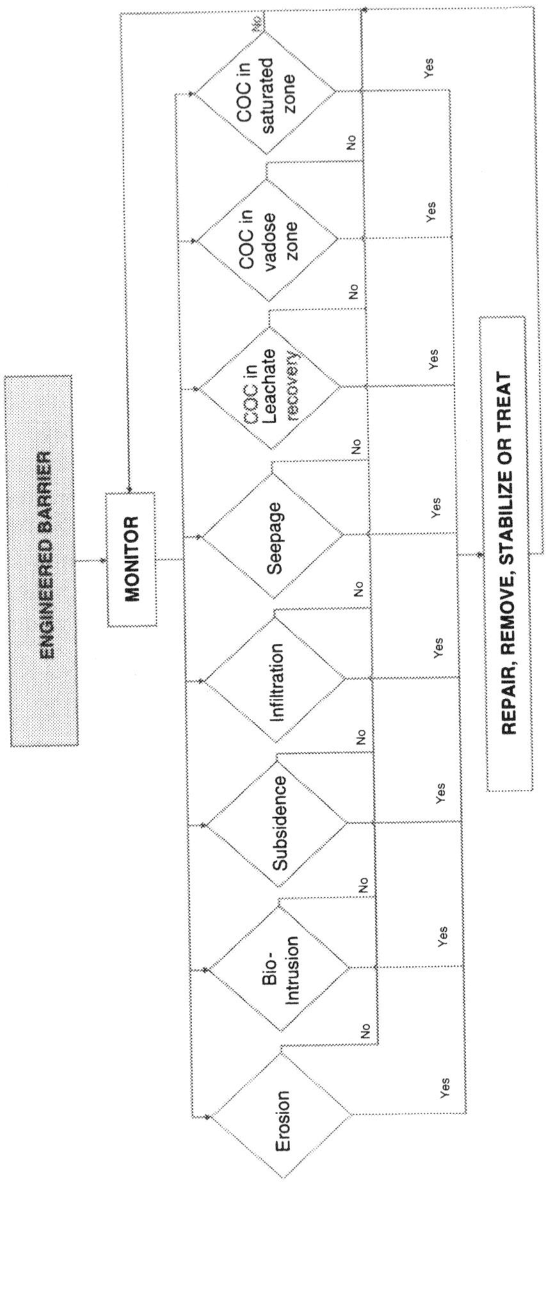

Fig. 4. Level 2 Logic Diagram for Engineered Barrier Maintenance (Sanchez et al., 2002).

site and all associated functions, maintain a multi-generational awareness within the local community, maintain emergency/contingency plans and performing emergency actions when applicable, maintain information/records, evaluate the surrounding environment/ecosystem, and continually assess the performance of the system and identify areas for improvement. Fig. 5 illustrates the Level 2 logic diagram for the Institutional Controls.

A major objective of many institutional controls is to keep people from inadvertently entering or ingressing a hazardous residual waste site. Conversely, the primary objective of most engineered barriers is to physically prevent contaminants from leaving or egressing the site. Appropriate site stewards must be identified to continually perform the aforementioned requirements to achieve these objectives.

Site stewards will need to communicate and coordinate with federal, state, and local regulators because the legal authority required to maintain and enforce these requirements is not always clearly defined and often depends on various statutes and regulations (Borinsky, 1995; Edwards, 2000a, 2000b; PaDEP, 2002). As such, if responsibilities and authority are not clearly defined, site management conditions will likely worsen over time and will impact maintenance and enforcement efforts.

EXTERNAL MECHANISMS

The management of residual contaminants does not always work as expected because the performance of each function and activity is continually influenced by external mechanisms, as illustrated in Fig. 6. External mechanisms, important to the sustainable management of residual contaminants, can be viewed in terms of environmental characteristics and social characteristics.

Environmental Characteristics

The environmental characteristics of a given site play a critical role with regard to the performance of each operational function and subsequent activity and therefore influence the long-term integrity of any contaminant isolation facility. Environmental interactions that influence the performance of a contaminant isolation facility include hydrologic, climatic, biologic, geologic, and chemical influences at the surface, within the contaminant isolation facility, in the vadose zone and in the saturated zone.

An example of environmental influences on contaminant isolation facilities can be observed at remediated uranium mill tailing sites located

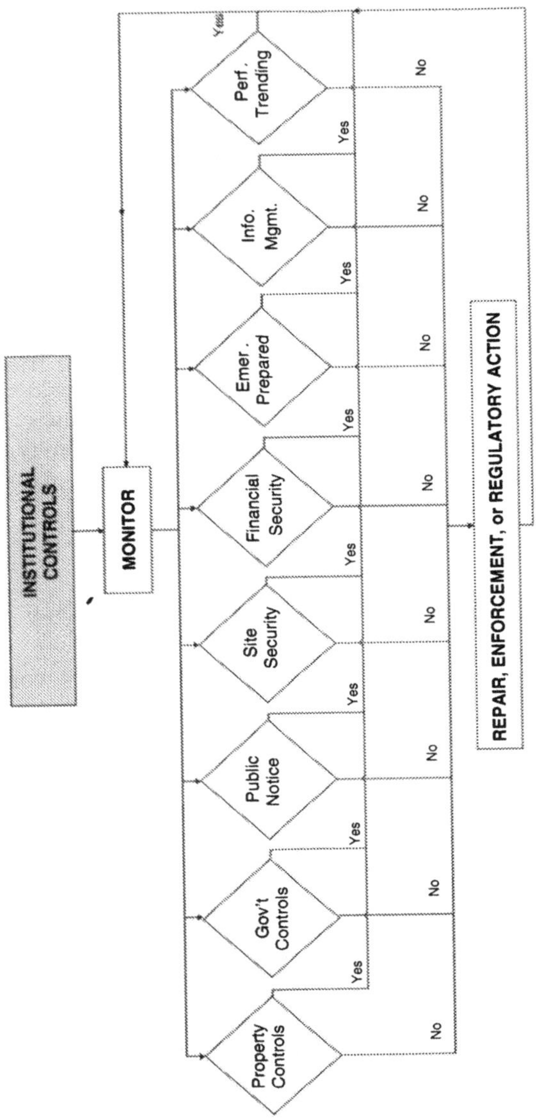

Fig. 5. Level 2 Logic Diagram for Institutional Controls.

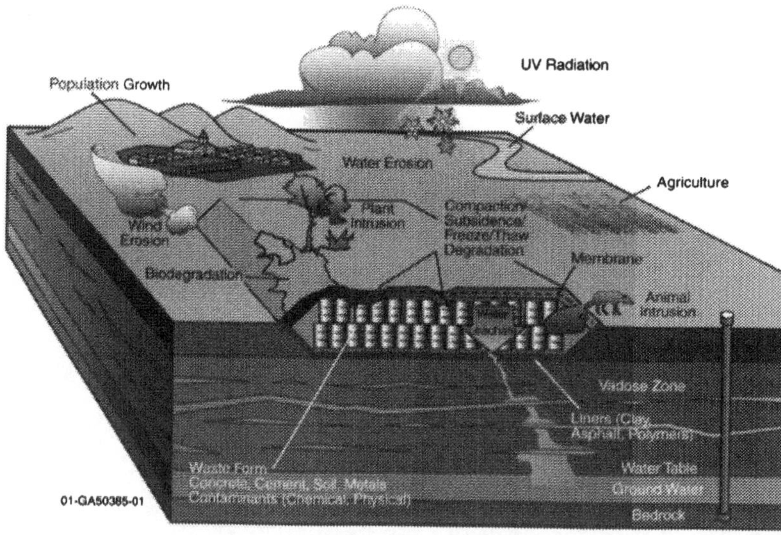

Fig. 6. Examples of External Mechanisms Influencing Contaminant Isolation (INEEL, 2001).

throughout the U.S. These sites were designed to properly function for 1000 years but at a minimum are required to last at least 200 years. Completed in the 1980s by the U.S. Department of Energy (USDOE), these sites provide the longest record of performance within our current regulatory framework.

The Burrell UMTRCA site in Pennsylvania illustrates how bio-intrusion can rapidly impact a surface barrier (Fig. 7). The Burrell isolation cell was completed in 1987 (USDOE, 2001). The cell cover consists of a 90 cm thick radon barrier of compacted soil, a 30 cm thick sand and gravel drainage layer, and a 30 cm thick rock layer (USDOE, 1997, 1999c). As early as 1988 observations of plant growth were reported on the cell cover. Within three years from construction, a diverse plant community was observed on the cap. Some tree species had rooted as deep as 33 cm into the compacted soil layer and an exotic perennial forb (Japanese Knotweed) had rooted 46 cm into the compacted soil layer (USDOE, 1997). Within 10 years from construction, plants were abundant over the entire cell cover.

The Durango UMTRCA site in Colorado illustrates how surface water run-off can erode surface barriers and potentially accelerate system degradation. Remedial actions at Durango, including the contaminant isolation cell, were completed in 1990 (USDOE, 2002). The top of the cell consists of

Fig. 7. Bio-intrusion at the Burrell, Pennsylvania, UMTRCA site (2003). Shrubs are growing in what was intended as a rocky cap, with the risk that their roots will penetrate to the buried wastes below.

a vegetated rock-and-soil matrix layer. The side slopes are covered with riprap and are steep in some places. Rock-armored drainage ditches lie at the bottoms of the side slopes on the northwest, south and east sides of the disposal cell. Immediately after the disposal cell was constructed, the basalt riprap drainage channels began to show signs of erosion as illustrated in Fig. 8. Significant erosion was observed at the mouth of the drainage channel situated on the east side of the cell. Within a few years after construction, the sandstone and shale that underlie the slopes above the ditches had weathered to small rubble that accumulated as colluvial deposits at the bottoms of the ditches (USDOE, 1999a).

DOE site stewards perform annual inspection and maintenance activities at each of these sites. As such, neither of these sites is believed to pose increased risk to human health or the environment at this time (USDOE, 2004a, 2004b, 2004c; Waugh, 2004). However, it is evident that the cover designs did not perform as originally anticipated.

Fig. 8. Surface erosion at the Durango, Colorado, UMTRCA site (2002).

Social Characteristics

Society can also influence the performance of contaminant isolation facilities. Isolation facilities and their engineered barriers are designed to contain the contaminants. In many instances the site stewards and regulators assume that individuals will stay away from these sites because of their knowledge of the associated hazards. However, changing social characteristics and multi-generational issues can challenge long-term isolation strategies.

Economic and demographic changes can influence regional land-use. Economic prosperity can encourage population expansion and subsequent encroachment toward once isolated facilities. This situation (i.e., new residential development toward once isolated facilities) can be observed at numerous USDOE sites, former U.S. Department of Defense (DOD) sites, as well at former chemical waste disposal sites such as the Love Canal site in Niagara Falls, New York (USDOE, 1999b).

Land-use changes from agricultural to industrial and residential complicate remedial decisions (McElfish, Pendergrass, & Mitchell, 1998; USDOE,

1998). Although it is relatively easy to consider current land-use during the remedy selection process, incorporating the future potential land-uses continues to be a challenging task (Wernstedt, Hersh, & Probst, 1998; Hersh et al., 2002).

An example of the importance of land-use changes with regard to institutional responsibilities can be observed at the American University site in Washington, DC (Fiala, 2001; Tucker, 2001). In support of the U.S. military efforts associated with World War I, the properties of the University were leased to the U.S. Army from 1917 to 1920. At the time, the site was a remote agricultural area. The U.S. Army and the U.S. Bureau of Mines used the property for the development and testing of chemical weapons. After World War I, the Army removed the temporary buildings, filled in the pits and trenches and returned the property to its "original landscape" as defined in the lease agreement. This agreement did not specify the need to reduce or eliminate the hazards. As the population of Washington, DC, continued to grow, the land-use of the area surrounding American University shifted from agricultural to residential. Portions of the original University property were sold for development and today some 1200 homes surround American University.

Approximately 70 years later the lack of institutional controls resulted in human intrusion into hazardous materials when, in 1993, chemical munitions were uncovered at a residential construction site (shown in Fig. 9). While this situation involved an historic decision to leave hazardous materials on-site, current regulations and techniques have not been able to prevent the problem of external intrusion. On two occasions, one in 1986 before the 1993 discovery and a second in 1995 subsequent to the discovery, the U.S. Army Corps of Engineers (USACE) evaluated the American University site and determined the site to be free from hazards (USGAO, 2002). Despite these assessments, the USACE estimated in 2002 that the cost of site remediation was over $125 million (USGAO, 2002).

Unanticipated social influences, such as regional demographic patterns and land-use shifts, will continually challenge remediation and post-closure long-term management strategies.

SUMMARY

A sustainable environmental protection system is required for the long-term management of residual contaminants. Such a system involves the integration of remedial processes, engineered barriers, and institutional controls.

Fig. 9. Environmental remediation at residential properties adjacent to American University site (2002).

Each of these subsystems includes required activities (i.e., institutional responsibilities) necessary for the system's continued performance.

Although site-specific characteristics are taken into account during the design phase of the system, these characteristics continually change and therefore influence the performance of the system throughout its life. Immediately after construction, social and environmental mechanisms begin to challenge the isolation facility. An adaptive management approach that can accommodate these temporal and spatial changes can improve the longevity of a contaminant isolation facility.

Key components of contaminant isolation facilities include surface and often subsurface-engineered barriers. While these barriers have been shown to be effective for the short term, their long-term performance remains questionable (USEPA, 1998). Recent operational experience suggests that current approaches may be inadequate given the timeframe that these materials must remain isolated. Environmental influences, such as hydrologic, climatic, biologic, geologic, and chemical influences have been shown to impact the performance of the contaminant isolation facility.

Additionally, management decisions made during the post-closure period appear key to the long-term success of the prescribed system. However, given that society has become more reliant on and confident of engineered structures, there appears to be a growing tendency to be even less concerned with these institutional aspects. Institutional responsibilities, including institutional controls and other required actions, appear to be a potentially critical route to system failure. Therefore, society should strive to strengthen and improve the monitoring, maintenance and accountability of institutional controls that are used for the isolation of residual contaminants.

REFERENCES

Benson, C. H., Albright, W. H., Roesler, A. C., & Abichou, T. (2002). Evaluation of final cover performance: Field data from the alternative cover assessment program. *Waste management*, Tucson, AZ: February, Waste Management Symposium, Inc.

Borinsky, S. C. (1995). The use of institutional controls in Superfund and similar state law. *Fordham Environmental Law Journal*, *7*(1995–1996), 54.

Breggin, L., Pendergrass, J., & McElfish, J. (1998). *Preliminary memorandum: Institutional controls over land uses at Superfund sites*. Environmental Law Institute, Washington, DC, March 2.

Brown, M. H. (1980). *Laying waste: The poisoning of America by toxic chemicals*. New York: Pantheon Books, New York.

CERCLA (Comprehensive Environmental Response, Compensation, and Liability Act). (1980). Public Law 96-150, as amended, 94 Stat. 2767, 42 USC 9601 et seq.

Clancy-Hepburn, M., Dycus, J., McElfish, J., Pendergrass, J., Spohr, D., Vaughn, D., & Wicke, H. (1995). *Institutional controls in use*. Environmental Law Institute, Washington, DC, Report #922042, September.

Clarke, J. H., MacDonell, M. M., Smith, E. D., Dunn, R. J., & Waugh, W. J. (2004). Engineered containment and control systems: Nurturing nature. *Risk Analysis*, *24*(3), 771–780.

DERTF. (1996). *Making institutional controls effective*. Defense Environmental Response Task Force. Report of the Future Land Use Working Group, Memphis, TN, September 19.

Edwards, A. L. (1997). Contaminated property: Heads up – Types of institutional controls. *ASTM Task Group Developing Guide on Institutional Controls*, *13*(7), 3.

Edwards, A. L. (2000a). Long-term enforcement and stewardship of institutional controls. *The Bureau of National Affairs Environmental Due Diligence Guide*, *95*(231), 1071–1078.

Edwards, A. L. (2000b). Raising the stakes: EPA region V's challenge regarding the long-term viability and enforceability of institutional controls at Brownfields Sites. American Bar Association's Trends.

ETF. (1998). Love Canal collection – The ecumenical task force of the Niagara frontier. The Ecumenical Task Force of the Niagara Frontier; University Archives, State University of New York Buffalo, October 8, 2002.

Fiala, C. J. (2001). *Congressional testimony – Military knowledge of buried munitions in Washington, DC. Committee on Government Reform, Subcommittee on the District of Columbia*. Washington, DC.

Gaspar, C., & Burik, D. V. (1998). *Local government use of institutional controls at contaminated sites*. Working Draft. International City/County Management Association, Washington, DC, January.

Gibbs, L. M. (1998). *Love Canal – The story continues*. State University of New York Press. ISBN# 0-86571-382-0.

Gibbs, L. M., & Levine, M. (1982). *Love Canal – My story*. New Society Publishers.

Hersh, R., Probst, K., Wernstedt, K., & Mazurek, J. (2002). *Linking land use and Superfund cleanups: Uncharted territory – Internet Edition*. Resources for the Future, Center for Risk Management. http://www.rff.org/Documents/RFF-RPT-landuse.pdf

INEEL (2001). *Technical baseline for the long-term stewardship national program*. Idaho National Engineering and Environmental Laboratory, Idaho Falls, INEEL/EXT-01-01133, September.

INEEL (2003). *Long-term stewardship science and technology roadmap*. Idaho National Engineering and Environmental Laboratory, Idaho Falls, ID, DOE/ID-10926, September.

Looney, B. B., & Falta, R. W. (2000). *Vadose zone science and technology solutions*. Columbus, OH: Battelle Press.

Mazur, A. (1998). *A hazardous inquiry – The Rashomon effect at Love Canal*. Harvard University Press.

McElfish, J. M., Pendergrass, J. A., & Mitchell, M. (1998). *Institutional controls case study: Mound plant*. Washington, DC: Environmental Law Institute ELI Project #941736, #972200.

NRC. (1997). *Barrier technologies for environmental management – Summary of a workshop*. Barrier Technologies for Environmental Management.

NRC. (2000). *Long-term institutional management of U.S. Department of Energy legacy waste site*. Committee on the Remediation of Buried and Tank Wastes, National Research Council, Washington, DC.

NYS. (1953). *Deed – Quit Claim between Hooker Electrochemical Company and The Board of Education of the School District of the City of Niagara Falls, New York*. New York State, Niagara County Clerk's Office, Niagara Falls, New York, April 28.

NYS. (1988). *Love Canal, Emergency Declaration Area, Decision on Habitability*. State of New York, Department of Health, Love Canal Technical Review Committee, Albany, NY, September.

PaDEP (2002). The Guardian Trust. Pennsylvania Department of Environmental Protection, July 22. Available at http://www.dep.state.pa.us/dep/deputate/airwaste/wm/landrecy/guardian.html

Pendergrass, J. (1999). Sustainable redevelopment of Brownfields: Using institutional controls to protect public health. *Environmental Law Reporter*, *29*(5–99), 16.

RCRA (1976). *Resource Conservation and Recovery Act*. US Code, Title 42, Chapter 82, Subchapter I, Section 6901, 42 U.S.C. 6901-6992; 90 Stat. 2795, October 21, 1976, as amended.

Russell, M. (2000). *Reducing the nuclear legacy burden: DOE environmental management strategy and implementation*. Joint Institute for Energy & Environment, Knoxville, April.

Sanchez, F., Clarke, J. H., & Parker, F. L. (2002). Evaluating requirements for stewardship of contaminated facilities. *Waste Management*, Tucson, AZ: WMSymposia, Inc.

Suthersan, S. S. (2002). *Natural and enhanced remediation systems*. Lewis Publishers.

Thompson, C. (2004). *Original Superfund site declared clean*.Niagara Falls. NY: Associated Press.

Tucker, J. (2001). Chemical weapons: Buried in the backyard. *Bulletin of the Atomic Scientists*. September/October, 51–56.

UMTRCA. (1978). *Uranium Mill Tailings Radiation Control Act*. Title 42, Chapter 88, Section 7901, November 8, 1978, as amended.

USDOD. (1998). *A guide to establishing institutional controls at closing military installations*. U.S. Department of Defense, February.

USDOE. (1997). *Effects of root intrusion at the Burrell, Pennsylvania Uranium Mill Tailings Disposal Site*. U.S. Department of Energy, Grand Junction, CO, GJO-97-5-TAR.

USDOE. (1998). *Effects of future land use assumptions on environmental restoration decision making*. U.S. Department of Energy, Office of Environmental Policy and Assistance, Washington, DC, DOE/EH-413/9810, July.

USDOE. (1999a). *Durango, Colorado disposal site*. Annual inspection report. U.S. Department of Energy, Long-Term Surveillance and Maintenance Program, Grand junction, CO.

USDOE. (1999b). *From cleanup to stewardship, a companion report to accelerating cleanup: Paths to closure and background information to support the scoping process required for the 1998 PEIS settlement study*. U.S. Department of Energy, DOE/EM-0466, October.

USDOE. (1999c). *Plant encroachment on the Burrell, Pennsylvania, Disposal cell: evaluation of long-term performance and risk*. U. S. Department of Energy, Grand Junction, CO, GJO-99-96-TAR, June.

USDOE. (2001). *Burrell, Pennsylvania, Disposal site, Long-term surveillance and maintenance program fact sheet*. U.S. Department of Energy, Long-Term Surveillance and Maintenance Program, Grand Junction, CO, July.

USDOE. (2002). *Environmental assessment of ground water compliance at the Durango, Colorado, UMTRA project site*. U.S. Department of Energy, Grand Junction, Colorado, DOE/EA-1452, November.

USDOE. (2004a). 2003 *Annual site inspection and monitoring report for Uranium Mill Tailings Radiation Control Act Title I disposal sites*. U.S. Department of Energy, Long-Term Surveillance and Maintenance Program, Grand Junction Office, Grand Junction, CO, GJO-2003-528-TAC, January.

USDOE. (2004b). 2003 *long-term surveillance and maintenance report*. U.S. Department of Energy, Office of Legacy Management, Grand Junction, CO, Grand Junction, CO, DOE-LM/GJ548-2004, July.

USDOE. (2004c). *LTSM Program 2003 UMTRCA Title I Annual Report, Burrell, Pennsylvania*. U.S. Department of Energy, Long-Term Surveillance and Maintenance Program, Grand Junction Program Office, Grand Junction, CO, January.

USEPA. (1988). *EPA Superfund record of decision: Love Canal*. U.S. Environmental Protection Agency, Office of Emergency and Remedial Response, Niagara Falls, NY, EPA/ROD/R02-88/063, September.

USEPA. (1991). *Love Canal records of decision*. U.S. Environmental Protection Agency, EPA/ROD/R02-85/014, EPA/ROD/R02-88/055, and EPA/ROD/R02-91/165.

USEPA. (1998). *Evaluation of subsurface engineered barriers at waste sites*. U.S. Environmental Protection Agency, Office of Solid Waste and Emergency Response, EPA-542-R-98-005, August.

USEPA. (2000). *Institutional controls: A site manager's guide to identifying, evaluating and selecting institutional controls at Superfund and RCRA corrective action cleanups*. U.S. Environmental Protection Agency, Washington, DC, EPA 540-F-00-005, September.

USEPA. (2002). *Small business liability relief and Brownfields revitalization act*. U.S. Environmental Protection Agency, March 7. Available at http://www.epa.gov/swerosps/bf/.

USEPA. (2003a). *Five-year review report of the Love Canal Superfund site*. U.S. Environmental Protection Agency,, Niagara Falls, New York, September.

USEPA. (2003b). *Love Canal summary – Region 2 Superfund*. U.S. Environmental Protectin Agency, July 31. Available at http://www.epa.gov/cgi-bin/epaprintonly.cgi.

USGAO. (1990). *Hazardous waste – Funding of postclosure liabilities remains uncertain*. U.S. General Accounting Office, Washington, DC, GAO/RCED-90-64, June.

USGAO. (2002). *Environmental contamination – Uncertainties continue to affect the progress of the Spring Valley cleanup*. Statement of David G. Wood, Director Natural Resources and Environment. U.S. General Accounting Office, Washington, DC, June 26.

Waugh, W. J. (2004). *Design, performance, and sustainability of engineered covers for Uranium Mill Tailings*. Workshop on Long-Term Performance Monitoring of Metals and Radionuclides in the Subsurface: Strategies, Tools, and Case Studies, U.S. Geological Survey, Reston, VA, April 21–22.

Wernstedt, K., Hersh, R., & Probst, K. (1998). Grounding hazardous waste cleanups: A promising remedy? *Land Use Policy, 16*, 45–55.

Whelan, E. (1985). *Toxic terror – The truth about the cancer scare*. Jameson Books.

Zuesse, E. (1981). "Love Canal – The truth seeps out". *Reason Magazine*, February, 16–33.

ECOLOGICAL CONSIDERATIONS FOR EVALUATING CURRENT RISK AND DESIGNING LONG-TERM STEWARDSHIP ON DEPARTMENT OF ENERGY LANDS

Joanna Burger, Nellie Tsipoura, Michael Gochfeld and Michael R. Greenberg

ABSTRACT

In this paper, we discuss methods to integrate ecological resources, ecosystem services, risk, and the transition to long-term stewardship on Department of Energy lands. Three types of information are required about ecological resources before decisions can be made about remediation, site transitions, and long-term stewardship: (1) the ecological resources and ecosystem functions (such as productivity) present on site and their spatial pattern, (2) the ecosystem services these resources provide to people, and (3) the risks from the interactions between people and these ecosystems. Once the ecological resources and ecosystem services are evaluated, then decisions about future land use, preservation, conservation, or protection of ecological resources within a designated land use can be implemented. Long-term stewardship requires both ecosystem

Long-Term Management of Contaminated Sites
Research in Social Problems and Public Policy, Volume 13, 139–162

ISSN: 0196-1152/doi:10.1016/S0196-1152(06)13006-9

protection in terms of biological resources and ecosystem function as well as biomonitoring to ensure minimal radiological or chemical risk and to inform future management. In some cases, protection of ecological resources may be preferable to cleanup that is physically disruptive, provided land use designation is consistent with ecological protection. In such instances, less site cleanup can prove preferable to more.

INTRODUCTION

Ensuring both human and ecological health is important to the public, managers, and policy-makers, particularly for contaminated lands. Several public and private agencies and organizations, such as the U.S. Department of Energy (DOE) and U.S. Department of Defense (DOD), have lands that contain chemical and radiological wastes that require remediation and restoration, or long-term stewardship to prevent risks to humans and ecological receptors. Herein we use ecological receptors to refer to biota, and to distinguish them from human receptors (workers, the public). The cleanup task on contaminated sites in the US is daunting in terms of amount of land, potential cleanup costs, and risks to humans and ecosystems. For example, the U.S. DOE has an environmental management task whose costs averaged about $6 billion a year over the last decade. Those costs represent 20% of the world's environmental remediation market (Sink & Frank, 1996) and nearly equal the entire budget of the U.S. Environmental Protection Agency. The Army is executive agent for the cleanup programs of the DOD, and the U.S. Army Corps of Engineers manages and executes the program (Lubbert & Chew, 2001). Further, neither DOE nor DOD currently owns all of the property it is responsible for cleaning up, complicating management, and future land use decisions. While some of these lands are contaminated to the extent that they pose a risk to humans or ecosystems, on others contamination is limited to small sections. Many contaminated sites are buffered by land that is not contaminated, is lightly contaminated, or is currently used for recreation, agriculture, or industry.

Managers, policy-makers and the public are faced with several decisions about contaminated lands that are not mutually exclusive: (1) how to reduce the current risk to humans and ecosystems (e.g., by remediation or isolation), (2) how to manage or remediate to reduce future risks (and future costs), and (3) how to ensure sustainability of lowered risks to human and ecological health. Protection of human and ecological health is the driver for current remediation/restoration, future land use, and long-term stewardship.

Options for the DOE, as shown in Fig. 1, include maintaining the status quo, conducting some level of remediation, and instituting alternative land uses ranging from preservation to industrial missions (DOE, 1999a).

Spatial information on ecological resources (both on DOE lands and on adjacent lands), ecosystem functioning, ecosystem services, and risk should inform selection among options. The DOE could retain ownership and management of the land itself, or reduce the physical size (or footprint) of their lands by transferring lands to private ownership. Retaining ownership where risk is low to workers and the public will require continued public support, perhaps in the context of a new or continuing mission. While footprint reduction was initially viewed as a preferred option for some DOE lands (Brown, 1998; Burger et al., 2003), we argue that designating portions of some DOE sites as "environmental parks" that will be maintained by the federal government in perpetuity is the optimal solution (Burger et al., 2003; Whicker et al., 2004). This is particularly important for some of the larger sites where there is a continued mission, where there is residual contamination, and where there are unique or rare ecosystems (Burger et al., 2003;

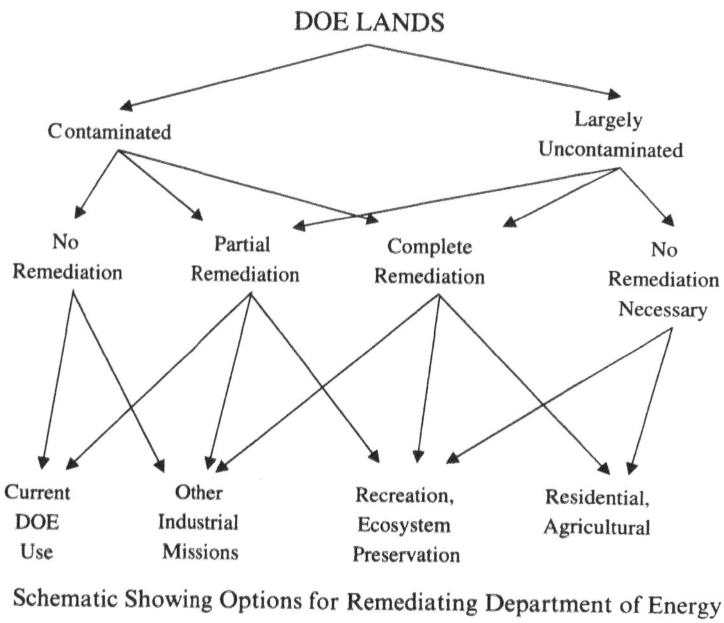

Fig. 1. Schematic Showing Options for Remediating Department of Energy Lands.

Whicker et al., 2004; DOE, 1994a; Nelson, 2001). Recently, the DOE reviewed its environmental management program and concluded that despite large expenditures of money, the cleanup tasks generally were not being completed (DOE, 2002). This resulted in a policy shift to link future land use and end-state planning with cleanup goals and residual risk (DOE, 2003). That is, cleanup levels should reflect projected future land use, what DOE terms a risk-based end state.

In this paper, we examine the issues surrounding protection of ecological health during the transition from the current state on DOE lands to long-term stewardship, and thereafter. Protecting ecological health now and in the future requires not only evaluating the risks to organisms and ecosystems but also understanding the value (or importance) of those ecological resources locally and nationally. While we use the DOE as a case study, the approaches and conclusions apply to other contaminated sites. We discuss the factors that will lead to a decision about protection of ecological resources, and how to ensure continued ecological health and well-being.

Management options for ensuring ecological health on contaminated lands require an integrated and iterative process to arrive at solutions. The process should involve adaptive management where future management actions are derived directly from assessment of current practices and risks. The adaptive management process should include a wide range of multidisciplinary expertise, agencies, and stakeholders, particularly for DOE sites with greater complexity, geographical size, and magnitude of the problems associated with chemical and radiological contamination. Multidisciplinary teams (e.g., managers and experts in remediation, public health, toxicology, ecology, land-use planning, economics, and risk communication) are faced with a multitude of inter-related challenges. These include addressing past contamination, identifying hazards and estimating risks of current contamination, choosing cost-effective environmental cleanup technologies, restoring habitats, determining appropriate future land-use options, and implementing long-term stewardship of lands that are not cleaned up to unrestricted use.

Management of contaminated lands has to date largely focused on immediate remediation, restoration, and risk reduction for humans (and in passing, ecological receptors), rather than on the transition to future land uses and long-term stewardship of lands that will retain some degree of contamination. While initially the public assumed that contaminated lands would be cleaned up to residential standards, it soon became evident that most sites would not be cleaned up to such a level, largely due to costs (DOE, 1999b; NRC, 2000). Cleaning up DOE sites to residential standards may also incur a health cost to both workers and the ecosystem. For many DOE sites, buffer lands have remained undisturbed for 50 years, and as a result there are valuable functioning ecosystems on site (Burger et al., 2003; Whicker et al., 2004; Dale & Parr, 1998; Malone, 1998). Ultimately, it is a public decision whether federal money should be spent for public ownership and re-use of these contaminated lands.

THE DEPARTMENT OF ENERGY CLEANUP TASK

The US nuclear weapons complex comprises approximately 5,000 facilities located at 16 major sites and more than 100 smaller sites (Crowley & Ahearne, 2002). Some 113 of the DOE sites around the country contain chemical and radiological wastes generated by the production of nuclear weapons (DOE, 2000). The contaminants include solvents, metals, and radionuclides in mixed waste; some are in underground storage tanks that currently have no proven technological method for cleanup. Many of the larger sites, such as Oak Ridge Reservation, Savannah River Site, and Hanford, have hundreds of individual waste sites within their boundaries that require remediation. In 1989, the DOE established an Office of Environmental Management (EM) to deal with the remediation tasks on its facilities (Sink & Frank, 1996; Daisey, 1998). Since 1994, EM's budget has averaged $6 billion a year in constant 1992 dollars (DOE, 2000; Frisch et al., 1998).

In general, cleanup at DOE sites will not have removed many of the long-lived radioactive and hazardous contaminants, necessitating long-term stewardship into the indefinite future (DOE, 1996a,b, 1999b). Some organics can be destroyed, but metals will remain for long periods of time. Radiological contaminants left on site will be stabilized and monitored and, where feasible, contained. Removal efforts for subsurface contaminants using current-day technologies will be very expensive and may disrupt ecosystems (NRC, 2000). On-site barriers (waste-site caps, surface barriers, and other engineered structures) that contain contaminants underground are less disruptive, but are difficult to monitor for possible failure (NRC, 2000). These DOE problems are not unique.

DOE cleanup goals differ markedly both across and within sites, including cleanup to residential standards, conversion to agricultural land or industrial facilities, use for recreational activities, and protection of natural areas. Cleanup to residential and agricultural-use standards can cost up to three times as much as cordoning off the site and relying on institutional controls for protection (DOE, 1995). Cleanup to accommodate other land uses (e.g., commercial/industrial or even recreational) would impose less stringent burdens, though at costs that are still significantly higher than those of the "iron fence" solution (DOE, 1996c). Past approaches for establishment of cleanup targets were primarily driven by human health protection. This paper focuses on protection of ecological health on DOE lands.

AN ECOLOGICAL FRAMEWORK FOR TRANSITION TO LONG-TERM STEWARDSHIP

Over the past 10 years, several authors have noted the value of ecological resources on DOE lands (Brown, 1998; Burger et al., 2003; Whicker et al., 2004). In this paper, we suggest an ecological framework for evaluating and protecting ecological resources that involves examining remediation decisions in terms of ecological impact, evaluating ecological resources and ecosystem services, instituting engineering and institutional measures to protect these resources, developing failure scenarios, and instituting biomonitoring to ensure early warning of possible ecological consequences of remediation of contaminants left in place (Table 1).

Biomonitoring refers to the regular, periodic monitoring of biota, measuring such things as reproductive success, behavioral changes, population levels, contaminant levels or effects, and health and well-being. Biomonitoring will provide information for adaptive management of DOE's contaminated lands.

From an ecological perspective, cleanup decisions that are disruptive of currently functioning ecosystems should address each of the following questions (Burger et al., 2003):

1. Is remediation necessary to protect ecological as well as human health, now or in the future?

Table 1. A Framework for Integrating Ecological Protection on Contaminated Department of Energy Lands.

Examining remediation decisions in an adaptive management framework
Evaluating ecological resources (from species to ecosystems)
Evaluating ecosystem services
Instituting engineering and institutional measures to protect ecological resources
Developing failure scenarios, and instituting adaptive management strategies to cope with failures
Instituting biomonitoring to ensure early warning of ecological consequences and inform future management

2. Are the benefits of reducing risks to humans sufficient to justify the risks to ecological receptors from remediation?
3. What are the most valuable DOE lands for their biodiversity, unique habitats, ecological services, or attributes?
4. What legal and institutional mechanisms exist to protect valuable ecological resources on DOE lands now and in the future?

Risk should be viewed within the context of possible harm to workers, the public, and ecosystems. Our working hypothesis, based on experience at individual DOE sites, is that with increasing levels of remediation there are increases in risks to workers and ecosystems, often with a less than offsetting diminution in risk to the general public (Fig. 2).

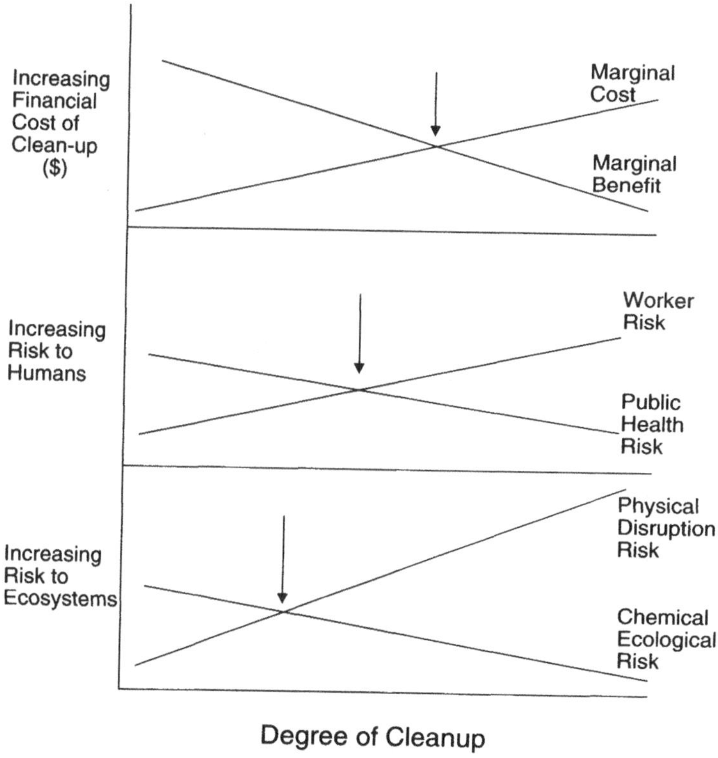

Fig. 2. Hypothetical Relationship between Levels of Remediation (from no Remediation on the Left, to Physical Disruption, Including Soil Removal on the Right) and Risk to Humans and Ecosystems.

In addition, from an ecosystem standpoint, the risk from physical disruption during remediation may exceed any reduction in risk due to chemicals. Integrating across both human and ecological risk, we are led to seek the point at which increased remediation incurs a greater cost to humans (workers) or ecosystems than the benefits to the public from the remediation. As portrayed in Fig. 2, cleanup to support residential use of a contaminated site comes at high financial cost (perhaps justified by projected human use benefits: top graph) but it may entail a level of cleanup that imposes too high worker risk (middle graph), and/or ecological risk (bottom graph) compared to the optimal level. The points (shown by arrows) where benefits to the public are less than the risks to workers, and where benefits of contaminant removal on ecosystems are less than the risks caused by physical disruption, should be determined and considered by the public and policy-makers.

We suggest that a framework for protecting ecological resources, from now into the future regardless of when remediation occurs, involves at least three aspects: (1) assessing ecological resources (biodiversity, unique habitats or species assemblages, ecosystem functioning), (2) assessing hazards (radiological, chemical, or physical risks), and (3) assessing ecosystem services in light of hazards. These aspects should be managed through four processes (identification and assessment, decisions about future land uses, risk reduction, and biomonitoring). All are required to adequately protect ecological resources. These important aspects of protecting ecological resources are further discussed in the section that follows.

The iterative processes to protect ecological resources include (1) identification and assessment of the ecological resources themselves (see evaluation considerations, below), (2) decisions about future land use with consideration of whether to preserve, conserve, or use the ecological resources (rather than permit their destruction during remediation), (3) reduction or avoidance of risks from radiological, chemical, and physical disruptions, and (4) biomonitoring. All four steps are essential, are iterative, and will need to have institutional controls to ensure that they continue. Initial identification and assessment of ecological resources can be a considerable task, but after such a baseline assessment has been completed, subsequent evaluations can be less time-consuming because they make use of this baseline information.

Reduction of risk includes addressing not only radiological/chemical contamination, but also physical disruptions caused by the original DOE missions, by remediation activities themselves (ranging from road construction to soil removal) and by constraints associated with future use, including infrastructure (e.g., roads built for recreation, forestry activities). For ecosystems that have been relatively undisturbed for 50 years, the physical disruptions caused by remediation and future use activities may pose a greater risk than the contaminants themselves.

Future land-use decisions are also an iterative process that involves DOE, regulators, and other stakeholders (Burger et al., 2003; DOE, 1996a, 2003). One option for DOE lands that are lightly contaminated, or not contaminated, and have valuable functioning ecosystems, is to give them legal protection as environmental parks (Burger et al., 2003; Nelson, 2001; Whicker et al., 2004). There is a basis for this in the designation of parts of several large DOE sites as National Environmental Research Parks (NERPs), following a Congressional direction to federal agencies (DOE, 1994b). In the 1970s, the Atomic Energy Commission, under the National Environmental Policy Act of 1969, began creating NERPs to recognize their ecological value (Brown, 1998; Burger, 2000a; DOE, 1994b). NERPs were designed to facilitate examination of the environmental consequences of human actions related to energy and weapons production, to explore methods to eliminate or minimize the adverse effects of energy and weapons, and to train people in environmental science (DOE, 1997). Such a designation, by Congress or some other "in perpetuity" mechanism, can prevent destruction of rich biological communities and valuable habitats, removal of topsoil, and degradation of wildlife populations. This offers the advantage of reduction in remediation costs, preservation of functioning ecosystems, and reductions in health risks to workers who would otherwise be engaged in on-site remediation, albeit at the potential cost of long-term risks associated with the contamination left in place.

Biomonitoring of all aspects of ecological resources is key to minimizing risks during cleanup and during long-term stewardship (see transition considerations below). It is a tool that could provide early warning if engineering and institutional controls break down, and can serve as the basis for further adaptive management actions. The biomonitoring costs, if lands are designated as environmental parks, will be lower than for other uses, such as residential, although some biomonitoring will still be required.

EVALUATING ECOLOGICAL RESOURCES

The term "ecological health" refers to the general health and well-being of biota, food webs, and ecosystem functions (i.e., productivity, energy transfer). Ecosystem services are the direct uses made of ecosystems by people, together with the other values provided including provision of clean water and air, and recreation. While we can argue about the definition of well-being (or how to measure it), the ends of the continuum are well understood (degraded systems versus those where the structure and functions are intact). For DOE lands, understanding exposure to radionuclides and chemicals to both humans and other biota is important to minimizing risk. For at least some sites, we argue, preserving ecosystem health should also be a goal. For these reasons, ecosystem value and ecosystem services should be evaluated in addition to radiological, chemical, and physical risks.

Below we discuss evaluating ecological resources and associated ecosystem services. We suggest that while truly understanding and valuing ecological resources is a very complex and difficult task, there are some useful indices (e.g., species diversity, reproductive success, index of biotic integrity) that can be easily and cost-effectively applied. Ecologists often make models so complex that screening for important ecological resources is obscured; yet, some models (particularly conceptual site models) will improve overall understanding.

Ecosystem Value

Even though some DOE lands are contaminated, the 2.54 million acres of DOE land comprise a reservoir of considerable ecological value (Brown, 1998). Approximately, 79% of this land has been undisturbed for the last 50 plus years because it served to buffer the nuclear production facilities (DOE, 2001). These undisturbed, mainly uncontaminated lands are a positive legacy of the Cold War, and should be recognized as such. The presence of radiation danger and national security concerns kept people out of these buffer lands, which now offer some of the finest and least disturbed plant and animal habitats in the United States (Brown, 1998; Burger et al., 2003; Burger, 2000a). Most DOE lands are not true wilderness areas (i.e., never disturbed), but instead are large tracts with minimal disturbance in recent decades. Many sites were exposed to human activity early in the 20th century (often agriculture or grazing), but they have been spared further disturbance since their designation as DOE lands. Recognition that DOE sites are home to rare and endangered species, unique environments, and minimally disturbed ecosystems has stimulated many calls for their protection (Burger et al., 2003; Whicker et al., 2004; Dale & Parr, 1998; Malone, 1998).

While one could argue that all undisturbed land is "important" for the integrity of local ecosystems, at some level the relative value of these lands with respect to rare, endangered, or unique assemblages of species or habitats must be evaluated (both with respect to local and regional or national ecosystems). Most authors, however, have dealt with only one or two sites, and not a range of sites. We created a data set of 25 DOE sites slated for remediation to explore the availability of data on species presence, habitats, and ecological value (Fig. 3).

Information was derived from DOE environmental assessment reports, from talking to managers on site, and from natural resource trustees (Burger et al., 2004).

Evaluation of the ecological value of any site requires information on the ecological resources that are present; at the very least, this should include information on biodiversity (species richness and evenness). Information on species richness (number of species) is especially critical for species that are listed as threatened or endangered, whether federally or on a state level (Niemela, 2000). While biodiversity refers to all species, in practice biodiversity indices are often measured using a particular taxon, such as birds (O'Connell, Jackson, & Brooks, 2000) or mammals (Sawicka-Kapusta & Zakrzewska, 1997). Unique species assemblages (such as neotropical migrants) and habitats (Carolina bays on the Savannah River Site in South Carolina, or the pine barrens of Brookhaven National Laboratory on Long Island) can also be evaluated. Another part of understanding the value of ecosystems is identifying critical ecosystems or habitats that are required for endangered or threatened species, species of special concern, or unique assemblages of species (Schroeder, 1983; Bartell, Campbell, & Burger, 2002). We suggest that a few key measures can be useful in determining ecological value, including species richness of key groups (e.g., birds, mammals), number of invasive species, habitat diversity, and unique species assemblages or habitats. In addition, it is essential to understand the exposure of ecological receptors to contaminants, and the risks these hazards pose. The paradigm used for assessing the risk to ecological receptors includes problem formulation, hazard identification and assessment, dose–response estimation and exposure assessment, and risk characterization (NRC, 1983, 1993). The difficulty at DOE sites is that spatial information on species distribution and/or contaminant distribution may not be available.

While information on federally endangered and threatened species was available for almost all 25 DOE sites, this information was less available for state listed species. Instead, we had to compare state endangered and threatened lists with the list of species from a site to derive the listed species on site. Table 2 summarizes information on threatened and endangered species for the 25 DOE sites, derived from the original data available from

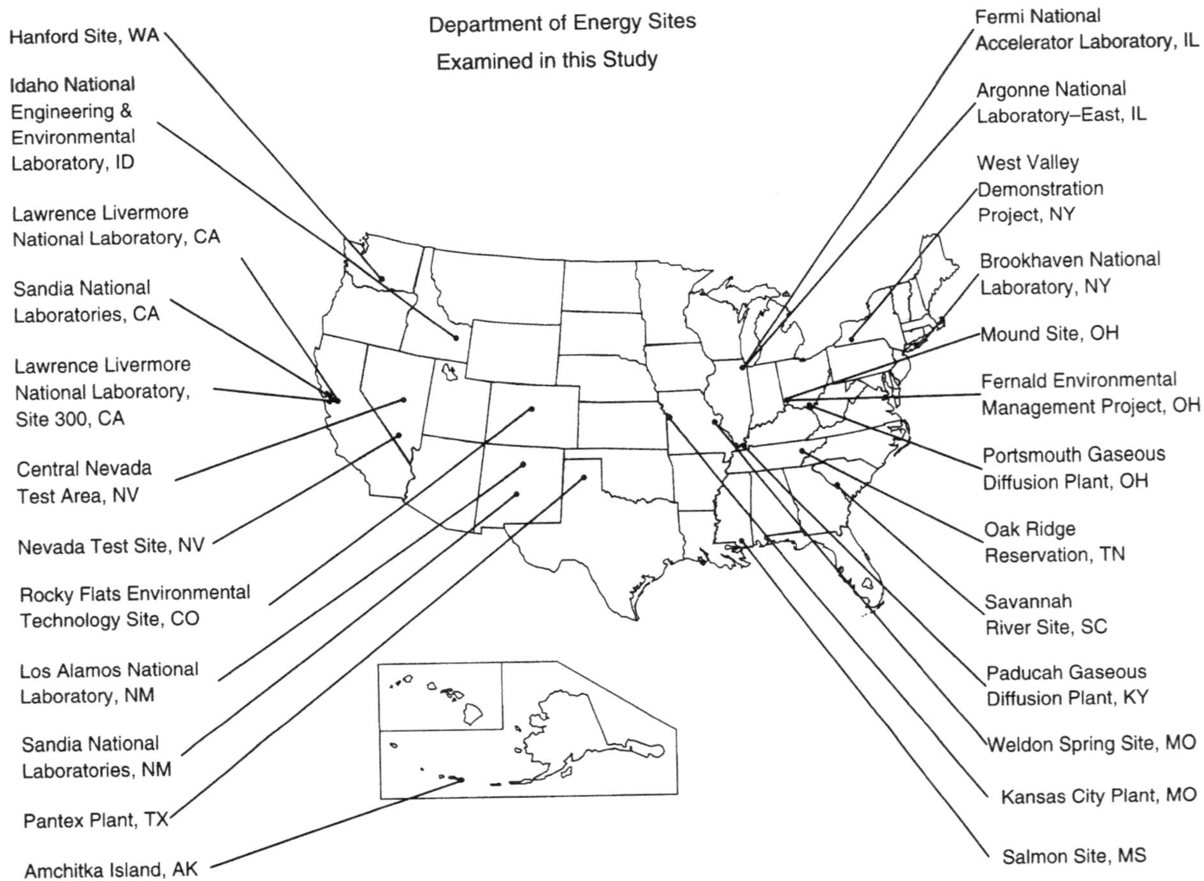

Fig. 3. Map Showing the Location of 25 DOE Sites where Ecological Resources were Evaluated.

Table 2. Department of Energy Sites Selected for Examination of Ecological Value and Number of Endangered and Threatened Species On Site.

Site Name	State	Size in Acres	Percent of Acres not Industrialized	Number Endangered and Threatened	
				Federal	State
Amchitka Island	AK	74,000	100	2	(a)
Argonne East	IL	1,500	87	0	6
Brookhaven	NY	5,263	67	0	10
Central Nevada Test Area	NV	2,560	100	1	1
Fermi Lab	IL	6,720	70	1	22
Fernald	OH	1,050	87	1	2
Hanford	WA	375,000	94	8	18
Idaho Engineering and Environmental Laboratory	ID	569,135	98	3	3
Kansas City Plant	MO	141	<10	0	0
Lawrence Livermore Main Site	CA	800	19	1	0
Lawrence Livermore Site 300	CA	7,067	95	3	2
Los Alamos	NM	28,000	(a)	5	11
Mound	OH	306	41	(a)	1
Nevada Test Site	NV	880,000	99	2	3
Oak Ridge	TN	35,000	70	2	16
Paducah	KY	3,423	78	0	2
Pantex	TX	10,177	76	1	6
Portsmouth	OH	3,714	96	0	2
Rocky Flats	CO	6,266	94	3	5
Salmon Site	MS	1,470	100	1	1
Sandia New Mexico	NM	2,820	67	0	1
Sandia California	CA	413	52	0	0
Savannah River	SC	198,344	90	7	13
Weldon Spring	MO	226	80	(a)	(a)
West V(a)lley	NY	230	13	(a)	4

Note: (a) Information not available or the site does not report this information.

DOE, and augmented by comparing the above lists. Taken altogether, the table indicates that there are important threatened or endangered species on DOE lands; 56% of the 25 sites had federal endangered or threatened species, and 76% had state endangered or threatened species.

Species richness information was most available for birds (80% of sites) and mammals (60%), less so for reptiles and amphibians (56 and 52%, respectively), and was almost nonexistent for invertebrates (8%). Information on invasive species was available for only 36% of the sites. The number of available habitats was directly related to total acres and nonindustrial acres. Species richness of birds, mammals, and reptiles (but not amphibians) was directly related to both total acres and total nonindustrial acres of sites (correlation coefficients of over 0.41, $p<0.01$). However, these data suggest that the area–size relationship explains only 17% of the variation in species richness – thus there is a clear need for continued biological assessment and monitoring on many of the DOE sites.

This information can be used in four key ways: (1) to assess the value of ecological resources on a given site, (2) to assess the value of ecological resources on DOE lands compared to adjacent lands, (3) to compare resources among DOE sites, and (4) to assess the potential of a given DOE site for bird-watchers, photographers, hunters, and others interested in recreation on site. If some lands on these DOE sites are declared environmental parks (or some other legally protected designation), then they could also be used for recreation, further serving the local community.

Ecosystem Services

Managing ecosystems is a social process (Norgaard, 1992; Meffe & Viederman, 1995), both in terms of protecting species and habitats and in terms of managing human uses and their effects on those ecosystems. Where land managers are required to make decisions about restoration and remediation as well as about current and future land uses understanding public attitudes and perceptions about how land should be used is essential (Lowrie & Greenberg, 1997). Such perceptions can not only aid in determining the required level of cleanup, but also can provide a path forward to these future land uses.

Burger and colleagues conducted a series of interviews at 3–6 events around each of four DOE sites shown in Fig. 3, ranging in size and population density: Brookhaven National Laboratory, Los Alamos National Laboratory, Savannah River Site, and Idaho National Engineering and Environmental Laboratory (Burger, 1997, 2000c, 2003, 2004; Burger et al., 1997; Burger et al., 1999a, 1999b, 2000a, 2000b). The protocol was to use the same questionnaire at all sites, administered by the same interviewers, using the same techniques. People at all events were interviewed while they waited for activities to begin, and refusal rates were less than 5% in each study. In all, 3,876 people were interviewed at 12 different events. Details of the individual studies can be found in the references mentioned earlier.

In one section of the surveys, subjects were asked to rate a series of potential future land uses; the list was selected from the future land use options which the Citizens Advisory Boards and DOE had previously determined for the sites (DOE, 1996a), and from activities that occurred locally and, in some cases, on the site. There was remarkable agreement among the perceptions and attitudes of the study populations (Burger, 2004), and the combined data set in Fig. 4 shows the overall ratings for the entire sample interviewed. In all cases, people rated bird-watching and maintaining the site as an NERP as the highest preferred land use, followed closely by hiking, camping, and other recreational activities. There was also general agreement in the uses that were rated the lowest, including building factories and homes, and storage of nuclear material. Even with such a diversity of people in widely scattered sites, there was general agreement on how these different lands should be used.

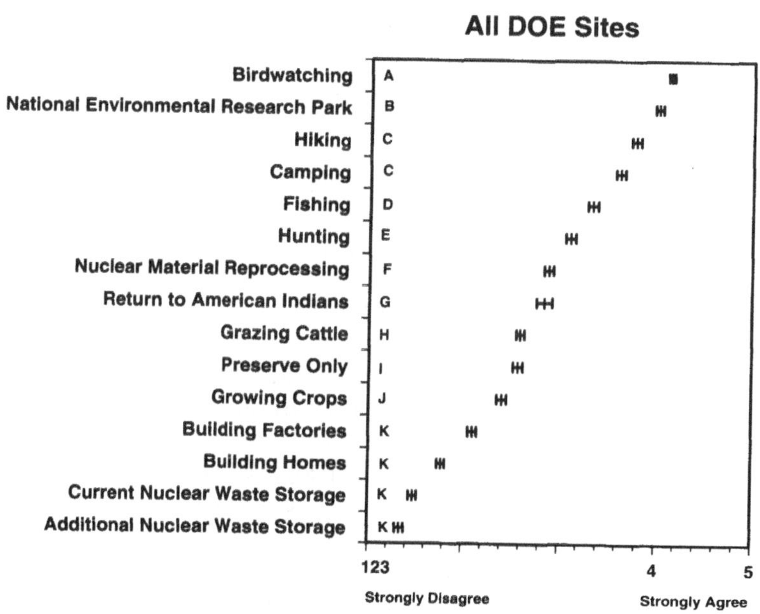

Fig. 4. Rating of Future Land Use Preferences for Four DOE Sites (Savannah River Site, Los Alamos National Laboratory, Idaho Engineering and Environmental Laboratory, Brookhaven National Laboratory).

While the residents living around these sites clearly perceived ecological value on the DOE lands, they also believed the lands should be used for recreation, and not just for preservation. This finding suggests that the risk to recreationists should be considered when determining cleanup levels, and when designing long-term stewardship. However, the potential conflicts between the isolation required by some endangered and threatened species and human use must also be considered. Biomonitoring schemes should be instituted to ensure public safety, and to provide early warning of any potential problems.

Ecological Resources, Ecosystem Services, and Risk

For the sample of 3,876 people above, a range of recreational activities was rated more highly than other land uses, which suggests that they or their friends would recreate on DOE lands if they were allowed. Some recreation, such as hunting is already allowed on small parcels of some DOE lands (Burger & Sanchez, 1999). The DOE has largely protected public exposure by providing buffers around their contaminated facilities and lands. In their Future Use Report, which involved stakeholders in the evaluation of possible future uses, DOE assumed that the average recreational user might be on their sites for a maximum of 14 days a year (DOE, 1996a).

Hunting and fishing rates were examined in the future land use studies described above (Burger, 1999a,b, 2000b,c, 2004). Between 30 and 40% of the people hunted, between 55 and 74% of the people fished, and between 35 and 71% of the people hiked. Some recreational activities were more popular at some sites than others: 58% fished at the Savannah River site (mean number of days 23 ± 1 day), 75% camped at Idaho National Engineering and Environmental Laboratory (mean of 14 ± 1 day), 71% hiked at Los Alamos National Laboratory (mean of 22 ± 2 days), and 74% fished at Brookhaven National Laboratory (mean of 38 ± 3 days) (Burger, 2004). Thus, the data clearly indicate that local people engage in an average of over 14 days for only one type of recreational activity, and many people engaged in several outdoor activities. While most people would not conduct all their recreation on DOE lands, some people could exceed the 14-day assumption (DOE, 1996a).

Above, we showed that DOE lands have valuable ecological resources, as indicated by the number of threatened and endangered species. Some of these DOE sites also have high numbers of resident and migrant birds, mammals, reptiles, and amphibians. These ecological resources can prove attractive to wildlife-watchers and photographers looking for suitable local places. Endangered/threatened species usually draw people who are interested in nature. For example, between 29 and 49% of the people interviewed said they birdwatched, and between 31 and 71% said they engaged in photography. The potential conflict between recreational uses (ranging from hunting to hiking) and protecting sensitive species must be considered, and such multiple uses might be managed by separating the uses spatially. For example, some endangered species may require isolation from human activities, while others may not.

We suggest that the combination of valuable ecological resources, high recreational rates, and preferred recreational uses for DOE lands might lead to the potential for DOE lands to be used by local residents if they were available. This suggests that during transitions, the potential risk to the public must be considered. We believe there are three key transition periods: (1) from the current state through remediation, (2) from current land use to future uses, and (3) from DOE ownership (or management) to other ownership. Any given DOE site might not experience all three of these. The risk during these transitions is not limited to workers and the public, but includes the ecological resources themselves because remediation can disrupt functioning ecosystems directly.

TRANSITION PROTECTION AND LONG-TERM STEWARDSHIP

While the majority of attention of the DOE has been directed at determining cleanup levels to protect human and ecological health, far less attention has been devoted to protecting ecological resources during remediation and following closure of sites where some contamination will remain. The initial and key step for protection of ecological resources during and after remediation is biomonitoring (Burger, 1999c). While biomonitoring should identify where engineering or institutional controls have failed (if contaminant levels rise in media or biota), it must also be borne in mind that there are institutional controls involved with monitoring that could also fail (NRC, 2000). Someone has to be in charge of biomonitoring in perpetuity.

Biomonitoring has three key aspects that could fail: obtaining reliable biological information, interpreting that information, and having a response mechanism to act on the information and interpretations. Obtaining biomonitoring information without interpretation and the ability to act will negate the purpose of the biomonitoring, which is to ensure continued ecological health. Some agent must be designated and funded to support these three aspects. There may be several ways to address this problem. One method may be to assign the first two tasks to a natural resource trustee and the last (further remediation or other action) to DOE, which retains the responsibility for radiological contamination on its sites. For example, at Amchitka Island where DOE's predecessor agency (the Atomic Energy Commission) set off three underground test shots, land ownership has been transferred to the U.S. Fish & Wildlife Service, but responsibility for the radionuclide contamination resides with DOE, which still has not reached closure agreements with state regulators and others. DOE has continuing responsibility, even after site closure.

Regardless of the level of remediation, biomonitoring is an essential tool for ensuring ecological health, providing a basis for adaptive management. Indicators, such as bioindicator species and biomarkers, are key components of biomonitoring schemes and should include indicators of both exposure and effects. Since it is not possible to monitor all species, interaction, and functions of ecosystems, the development of both site specific and complex-wide bioindicators is critical (Burger & Gochfeld, 2001, 2004; Carignan & Villard, 2001; Peakall, 1992; Piotrowski, 1985). Most books devoted to ecological risk assessment provide methods for evaluation at different levels of ecological organization, but do not provide a comprehensive plan for any one habitat or land type, although some have provided plans for regions (Cairns, 1990; Hunsaker, Carpenter, & Messer, 1990; Suter, 1990, 1993). Excellent methods are available (Linhurst et al., 1995), but authors seldom commit to a specific plan. Monitoring plans should take into account both the value and vulnerability of the ecosystems as well as the relative susceptibility of these ecosystems to radiological/chemical risks (Burger, 1997).

Monitoring schemes will be most useful if (1) they include many species representing different trophic levels, (2) indicator selection is based on sound quantitative databases, and (3) caution is used in interpreting population trends, contaminant levels, and other parameters (Peakall, 1992; Burger & Gochfeld, 2001, 2004; Carignan & Villard, 2001). A number of tools are available to aid in developing monitoring schemes, including sophisticated toxicity tests, remote sensing, geographic information systems (GIS), conceptual site models, and spatially explicit simulation models (Aspinall & Pearson, 2000; Cairns & Niederlehner, 1992, 1996). GIS tools are now available not only to show the spatial and temporal patterns of specific indicators, bioindicators, and biomarkers, but also to show relationships between different types of indicators. The tools, however, should not dictate the choice of indicators or biomarkers and in most cases suites of indicators will be required (Harwell & Kelly, 1990). There are generalized models for indicators at DOE sites as well as general characteristics of bioindicators applicable to DOE sites (Burger, 1999c; Burger & Gochfeld, 2001, 2004; Carignan & Villard, 2001).

Our point here is not to describe or define a biomonitoring plan for DOE, but to suggest that it is an essential part of ecosystem protection during and following cleanup, and is critical for sustainability and long-term stewardship. Further, institutional controls are essential for maintenance, interpretation, and action on that biomonitoring plan. A biomonitoring plan thus is not only part of institutional controls, but relies on other institutional controls for its continuation and efficacy.

CONCLUSIONS

DOE, and other agencies with large contaminated sites, must ensure both human and ecological health, both during remediation and in the future. Managers, policy-makers, and the public are faced with several decisions about contaminated lands that are not mutually exclusive: (1) how to reduce the current risk to humans and ecosystems (i.e., by remediation, isolation, or institutional controls), (2) how to manage or remediate to reduce future risks (and future costs), (3) how to achieve sustainability of ecosystems and decreased health risks, and (4) how to reduce land holdings (as mandated or encouraged under the current regulatory and institutional framework), while still protecting human and ecological health. Often, ecological health (and indeed worker health) is best protected by not disrupting by remediation ecosystems that are functioning and hold valuable resources (Burger et al., 2003). Options for the DOE include maintaining the status quo, conducting some level of remediation, and instituting alternative land uses ranging from preservation to industrial. The DOE could retain ownership and management of the land itself, or reduce the physical size of its lands. While footprint reduction was initially viewed as a preferred option for some DOE lands, we argue that designating portions of some DOE sites as "environmental parks" that will be maintained by the federal government in perpetuity is the optimal solution (Whicker et al., 2004). While this may reduce remediation costs and decrease disruption to ecosystems, it does not obviate the need for ecological assessment, decision-making about remediation and risk reduction, or biomonitoring within a long-term stewardship program.

Regardless of the future land use for a site, we suggest that there are ecological resources to consider during and after remediation, particularly following closure. These considerations include identification and assessment of ecological resources, decisions about remediation and management, risk reduction or natural attenuation, and biomonitoring of ecological resources. Biomonitoring is both a part of the institutional controls that should be in place for long-term stewardship and itself subject to institutional controls. Controls must be in place to conduct the biomonitoring, analyze the biomonitoring data, and allow for corrective action if the biomonitoring indicates a failure in the institutional or engineering controls to protect ecological health. We suggest DOE devote more time and resources to planning for closure and long-term stewardship of ecological resources on key DOE sites, and not just to the human health aspects of long-term stewardship.

ACKNOWLEDGMENTS

We thank K. Lowrie, T. Miller, and M.A. Carletta for help with the original analysis of 25 DOE sites, C. Dixon, R. Ramos, S. Shukla, C. Jeitner, K.F. Gaines, S. Boring, C. Lord, D. Rousch, and O. Myers for help with the perceptions and recreational rates research, the DOE sites for the information provided in phone interviews, and individual state agencies for providing information on threatened and endangered species. Over the years, our thinking about land use, risk, and end states has been influenced by C.W. Powers, B.D. Goldstein, D. Geiser, H. Mayer, J. Kang, J. Lehr, B. Friedlander, and D. Kosson. This research was funded by the Consortium for Risk Evaluation with Stakeholder Participation (CRESP) through the Department of Energy (AI # DE-FC01-95EW55084, DE-FG 26-00NT 40938). JB and MG were also partially supported by NIEHS grant ESO 5022. The results, conclusions, and interpretations reported herein are the sole responsibility of the authors, and should not in any way be interpreted as representing the views of the funding agencies.

REFERENCES

Aspinall, R., & Pearson, D. (2000). Integrated geographical assessment of environmental contamination in watch catchments: Linking landscape ecology, environmental modeling and GIS. *Journal of Environmental Management*, *59*, 299–319.

Bartell, S. M., Campbell, K. R., & Burger, J. (2002). Assessing ecological risks at U.S. Department of Energy facilities using methods borrowed from landscape ecology and habitat suitability analysis. II. Adapting Habitat Suitability Index models and landscape-scale ecological models for use in ecological risk assessment. *International Journal of Environment and Pollution, 2*, 52–77.

Brown, K. S. (1998). The great DOE land rush? *Science, 282*, 616–617.

Burger, J. (1997). Method for and approaches to evaluating susceptibility of ecological systems to hazardous chemicals. *Environmental Health Perspectives, 105*, 843–848.

Burger, J. (1999a). American Indians, hunting and fishing rates, risk, and the Idaho National Engineering and Environmental Laboratory. *Environmental Research, 80*, 317–329.

Burger, J. (1999b). Recreation, consumption of wild game, risk, and the Department of Energy sites: Perceptions of people attending the Lewiston (Idaho) "Roundup". *Journal of Toxicology and Environmental Health, 56*, 221–234.

Burger, J. (1999c). Environmental monitoring on Department of Energy lands: The need for a holistic plan. *Strategic Environmental Management, 1*, 351–367.

Burger, J. (2000a). Contaminated Department of Energy facilities and ecosystems: Weighing the ecological risks. *Journal of Toxicology and Environmental Health, 61*(2), 141–154.

Burger, J. (2000b). Recreation and risk around Los Alamos: Are Hispanics more at risk? *Journal of Toxicology of Environmental Health, 61*, 265–280.

Burger, J. (2000c). A comparison of on-site hunters, sportsmen and the general public about recreational rates and future land use for the Savannah River Site. *Journal of Environmental Planning and Management, 43*, 221–233.

Burger, J. (2003). Recreational rates and future land use preferences about Idaho National Engineering and Environmental Laboratory. *Journal of Environmental Planning and Management, 46*, 857–874.

Burger, J. (2004). Recreational rates and future land-use preferences about four Department of Energy sites: Consistency despite demographic and geographical differences. *Environmental Research, 95*(2), 215–223.

Burger, J., Carletta, M. A., Lowrie, K., Miller, K. T., & Greenberg, M. (2004). Assessing ecological resources for remediation and future land uses on contaminated lands. *Environmental Management, 34*, 1–10.

Burger, J., & Gochfeld, M. (2001). On developing bioindicators for human and ecological health. *Environmental Monitoring and Assessment, 66*, 23–46.

Burger, J., & Gochfeld, M. (2004). Bioindicators for assessing human and ecological health. In: G. B. Wiersma (Ed.), *Environmental monitoring* (pp. 541–566). Boca Raton, FL: CRC Press.

Burger, J., Leschine, T. M., Greenberg, M., Karr, J., Gochfeld, M., & Powers, C. W. (2003). Shifting priorities at the Department of Energy's bomb factories: Protecting human and ecological health. *Environmental Management, 31*, 157–167.

Burger, J., Roush, D., Ramos, R., & Gochfeld, M. (2000a). Risk concerns, land use, stewardship, and the Idaho National Engineering and Environmental Laboratory: Attitudes of the Shoshone-Bannock and other American Indians. *Environmental Research, 83*, 298–310.

Burger, J., Roush, D. E., Jr., Sanchez, J., Ondrof, J., Ramos, R., McMahon, M. J., & Gochfeld, M. (2000b). Attitudes and perceptions about ecological resources, hazards, and future land use of people living near the Idaho National Engineering and Environmental Laboratory. *Environmental Monitoring and Assessment, 60*, 145–161.

Burger, J., Roush, D. E., Jr., Wartenberg, D., & Gochfeld, M. (1999a). Future land use and concerns about the Idaho National Engineering and Environmental Laboratory: Perceptions and concerns of urban dwellers. *Environmental Management, 24*, 541–551.

Burger, J., & Sanchez, J. (1999). Perceptions of on-site hunters: Environmental concerns, future land use, and cleanup options at the Savannah River Site. *Journal of Toxicology and Environmental Health, 56*, 267–281.

Burger, J., Sanchez, J., Gibbons, J. W., & Gochfeld, M. (1997). Risk perception, federal spending, and the Savannah River Site: Attitudes of hunters and fishermen. *Risk Analysis, 17*, 313–320.

Burger, J., Sanchez, J., Gibbons, J. W., Ondrof, J., Ramos, R., McMahon, M. J., Gaines, K. F., Lord, C., Fulmer, M., & Gochfeld, M. (1999b). Attitudes and perceptions about ecological resources and hazards of people living around the Savannah River Site. *Environmental Monitoring and Assessment, 57*, 195–211.

Cairns, J. Jr. (1990). The genesis of biomonitoring in aquatic ecosystems. *Environmental Professional, 12*, 169–176.

Cairns, J. Jr., & Niederlehner, B. R. (1992). Predicting ecosystem risk: Genesis and future needs. In: J. Cairns, Jr., B. R. Niederlehner & D. R. Orvos (Eds), *Predicting ecosystem risk* (pp. 327–344). Princeton, NJ: Princeton Scientific Publishing Co.

Cairns, J. Jr., & Niederlehner, B. R. (1996). Developing a field of landscape ecotoxicology. *Ecological Application, 6*, 780–796.

Carignan, V., & Villard, M. A. (2001). Selecting indicator species to monitor ecological integrity: A review. *Environmental Monitoring and Assessment, 78*, 45–61.

Crowley, K. D., & Ahearne, J. F. (2002). Managing the environmental legacy of U.S. nuclear-weapons production. *American Scientist, 90*, 514–523.

Daisey, J. M. (1998). A report on the workshop on improving exposure analysis for DOE sites, September, 1996, San Francisco, CA. *Journal of Exposure Analysis and Environmental Epidemiology, 8*, 3–8.

Dale, V. H., & Parr, P. D. (1998). Preserving DOE's research parks. *Issues in Science and Technology, 14*, 73–77.

Department of Energy (DOE). (1994a). *Stewards of national resources*. Washington, DC: Department of Energy. DOE/FM-0002.

DOE. (1994b). *National environmental research parks*. Washington, DC: Department of Energy, Office of Energy Research. DOE/EM-0232.

DOE. (1995). *Estimating the cold war mortgage: The 1995 baseline environmental management report*. Washington, DC: Office of Environmental Management. DOE/EM-0232.

DOE. (1996a). *Charting the course: the future use report*. Washington, DC: Department of Energy. DOE/EM–0283.

DOE. (1996b). *Hanford site: Comprehensive facility and land use plan*. Washington, DC: Department of Energy. DOE/EM–0283.

DOE. (1996c). *Idaho national engineering laboratory: Comprehensive facility and land use plan*. Washington, DC: Department of Energy. DOE/EM–0283.

DOE. (1997). *Cross-cut guidance on environmental requirements for DOE real property transfers*. Department of Energy, Office of Environmental Policy and Assistance (October 1997, DOE/EH412/9712, p. 3–3), Washington, DC.

DOE. (1999a). *Nuclear and environmental stewardship dominate DOE spending*. Available at http://nmsnet.ilnl.gov/demo/demo/ndom.html

DOE, (1999b). *From cleanup to stewardship: A companion report to accelerating cleanup: Paths to closure*. Washington, DC: Office of Environmental Management. DOE/EM–0466.

DOE, (2000). *Paths to closure: Status report 2000*. Washington, DC: Office of Environmental Management. DOE-EM–0526.

DOE. (2001). *Long-term stewardship report to Congress.* Prepared to fulfill a requirement in the FY 2000 National Defense Authorization Act (NDAA). Department of Energy, Washington, DC.

DOE. (2002). *A review of the environmental management program.* Department of Energy, Washington, DC. Available at http://web.em.doe.gov/ttbr.pdf

DOE. (2003). *Use of risk-based end states.* Department of Energy, Washington, DC. (Policy 455.1, July 2003).

Frisch, M., Solitare, L., Greenberg, M., & Lowrie, K. (1998). Regional economic benefits of environmental management at the US Department of Energy's major nuclear weapons sites. *Journal of Environmental Management, 54,* 23–37.

Harwell, M. A., & Kelly, J. R. (1990). Indicators of ecosystem recovery. *Environmental Management, 14,* 527–545.

Hunsaker, C., Carpenter, D., & Messer, J. (1990). Ecological indicators for regional monitoring. *Bulletin of the Ecological Society of America, 71,* 165–172.

Linthurst, R. A., Bourdeau, P., & Tardiff, R. G. (1995). *Methods to assess the effects of chemicals on ecosystems.* Chichester, UK: Wiley.

Lowrie, K., & Greenberg, M. (1997). Placing future land use planning in a regional context: The Savannah River Site. *Federal Facilities Environmental Journal, 1,* 51–65.

Lubbert, R. F., & Chu, T. J. (2001). Challenges to cleaning up formerly used defense sites in the twenty-first century. *Remediation, 11,* 19–31.

Malone, C. R. (1998). Implications of resources management at the Nevada Test Site. *Federal Facilities Environmental Journal, 9,* 51–62.

Meffe, G. K., & Viederman, S. (1995). Combining science and policy in conservation biology. *Wildlife Society Bulletin, 23,* 327–332.

National Research Council (NRC). (1983). *Risk assessment in the federal government: Managing the process.* Washington, DC: National Academy Press.

NRC. (1993). *Issues in risk assessment.* Washington, DC: National Academy Press.

NRC. (2000). *Long-term institutional management of U.S. Department of Energy legacy waste sites.* Washington, DC: National Academy Press.

Nelson, R. H. (2001). *From waste to wilderness: Maintaining biodiversity on nuclear-bomb-building sites.* Competitive Enterprise Institute. Available at hhtp//www.cei.org/MonoReader.asp?ID=1411.

Niemela, J. (2000). Biodiversity monitoring for decision-making. *Annals Zoologica Fennici, 37,* 307–317.

Norgaard, R. B. (1992). Environmental science as a social process. *Environmental Monitoring and Assessment, 20,* 95–110.

O'Connell, T. J., Jackson, L. E., & Brooks, R. P. (2000). Bird guilds as indicators of ecological condition in the central Appalachians. *Ecological Applications, 10,* 1706–1721.

Peakall, D. (1992). *Animal biomarkers as pollution indicators.* London: Chapman and Hall.

Piotrowski, J. K. (1985). Individual exposure and biological monitoring. In: V. B. Vouk, G. C. Butler, D. G. Hoel & D. B. Peakell (Eds), *Methods for estimating risk of chemical injury: Human and non-human biota and ecosystem* (pp. 123–135). Chichester, UK: Wiley.

Sink, C. H., & Frank, C. W. (1996). DOE forges partnerships for environmental cleanup. *Forum for Applied Research and Public Policy, 11,* 65–69.

Sawicka-Kapusta, K., & Zakrzewska, M. (1997). Use of small mammals for monitoring heavy metal contamination in the environment. *Archives of Environmental Contamination and Toxicology, 33,* 172–181.

Schroeder, R. L. (1983). *Habitat suitability index models: red-spotted newt*. US Fish & Wildlife Service Biol. Rep. 82 (10.111). US Department of Interior, Washington, DC.

Suter, G. W., II. (1990). Endpoints for regional ecological risk assessment. *Environmental Management*, *14*, 9–23.

Suter, G. W., III. (Ed.) (1993). *Ecological risk assessment*. Boca Raton, FL: Lewis Publishers.

Whicker, F. W., Hinton, T. G., MacDonell, M. M., Pinder III, J. E., & Habegger, L. J. (2004). Avoiding destructive remediation at DOE sites. *Science, 303,* (12 Mar), 1615–1616.

THE LESSONS OF HISTORY

LIVING IN A NUCLEAR LANDSCAPE: REHABILITATION AND RESETTLEMENT OF PROVING GROUNDS IN AUSTRALIA AND ISLANDS OF THE WESTERN PACIFIC

Anne Ballou Jennings, Amy M. Seward and Thomas M. Leschine

ABSTRACT

Between 1946 and 1962 the United States detonated 109 nuclear weapons in an area of the Pacific Ocean called the Pacific Proving Grounds, mainly at the Enewetak and Bikini Atolls in the Republic of the Marshall Islands. The British nuclear testing program spanned eleven years (1952-1963) and involved detonation of 21 weapons in addition to a number of smaller-scale experiments (the "minor" trials) in South Australia. In both cases lands that were inhabited by an original people were extensively contaminated by the testing and, following its cessation, these people desired the return of their lands in a condition that would support resumption of traditional lifestyles. Cleanup and reoccupation of the

Long-Term Management of Contaminated Sites
Research in Social Problems and Public Policy, Volume 13, 165–192

ISSN: 0196-1152/doi:10.1016/S0196-1152(06)13007-0

contaminated lands in these two arenas raised similar questions of cultural identity and communication and the importance of cultural as well as technical understanding in establishing conditions for safe resettlement. The approaches taken to resolving these questions in the two situations have been very different—featuring extensive stakeholder involvement in a technocratic decision making process in the Maralinga case and a series of attempted negotiations and claims under the terms of trust agreements with the U.S. government in the Marshall Islands case. Similar questions regarding what constitutes a safe final condition and the means by which safety will be maintained remain in each case however. The Australian case has reached the more definitive outcome of the two. A jointly drafted comprehensive environmental management plan, if fully implemented and sustained over the time periods necessary, could facilitate a semblance of the outstation lifestyle that the aboriginal Maralingan people desire. But what has been agreed to may still prove very difficult to live with given the levels of residual contamination and the many ways that traditional cultural practices put people in intimate contact with the land.

INTRODUCTION

The U.S. Department of Energy (USDOE) and the Australian Nuclear Science and Technology Organisation (ANSTO), as well as a host of other agencies, have spent over fifty years and hundreds of millions of dollars on the rehabilitation of former nuclear weapons test sites in Australia and the Pacific that they hope to turn over to traditional landowners.[1] Topsoil and debris at the sites are contaminated with cesium, strontium, americium, and plutonium and the lands' traditional owners are demanding cleanup.

While there are many issues with which regulators are wrestling, much of the conflict and controversy stems from these questions: How clean is clean enough, given that human habitation is the goal? Who gets to decide? How will cross-cultural communication be developed and maintained in the name of managing the remaining risks?

Those in charge of the cleanups are looking to national and international radiation protection standards for guidance, while traditional owners are looking at past, present and future land use and settlement patterns, and further express the desire for compensation to address human health and social impacts issues.[2] Marshall Islanders who left their traditional homes in the 1940s during U.S. nuclear weapons testing at Bikini and Enewetak and who were moved numerous times[3] want the islands restored to a condition that would permit unrestricted access and resettlement. The Maralinga Aborigines of South Australia, who were moved to mission outstations beginning in the 1920s and were later restricted from their traditional lands for British nuclear weapons tests, appear to be settling for a partial cleanup of the test site's 3,200 square kilometers of contaminated land.

Relations between governmental organizations and native populations—especially between U.S. authorities and the Marshall Islanders—have been strained over the years by a fundamental lack of trust. Reaching successful–and final–outcomes in these cases perhaps depends as much on rebuilding that trust as it does on meeting technical criteria. Applying national and international exposure standards—by necessity the focus of remediation strategies—has not been as straightforward as it might seem however. Complex social and cultural issues that underlie the situations in the Marshall Islands and the Maralinga lands seem to demand comprehensive approaches. Understanding site history and the culture of the displaced inhabitants, communicating risk, and demonstrating a commitment to resolving differences in scientific opinions have been important to the formulation of rehabilitation and resettlement plans. Stakeholder participation has also been critical to establishing the importance of land ownership, addressing long-term resource-use issues, and overcoming obstacles created by decades of mistrust. An important difference between the two situations however appears to be the extent to which the Marshallese were themselves exposed to fallout and other potentially harmful effects of testing, and the political salience that fact continues to have today. Questions of reparations for human health and other effects of testing appear to have merged with questions of coming to terms on mutually agreed site conditions that would permit more extensive resettlement and reuse of former test sites than has occurred to date, underscoring questions of trust.

Approaching the problem from a historical perspective gives the situation context by helping to establish the significance of land and land ownership, the significance of place, and the long-term impact of both relocation and contamination as a result of nuclear testing. This approach gives the traditional owners a voice and validates their concerns while also giving decision-makers perspective on the social impact of the loss of land as a result of contamination. A historical perspective can also serve to shift the focus of remediation planning at these sites from technical feasibility to cultural appropriateness and social justice while "sparking the arc between the social and the ecological" (Buell, 1998).

The three sites considered include:

Bikini Atoll, Republic of the Marshall Islands (RMI). Twenty-three nuclear weapons tests were conducted at Bikini Atoll, starting with *Operation Crossroads* in 1946 and ending with the Juniper test in 1958. The yield at Bikini exceeded seventy-five thousand kilotons and comprised 72% of the Pacific Testing Program's total yield. Bikini was declared safe in 1968 after a preliminary cleanup and was partially resettled in 1973. Human health monitoring revealed excessive radiation contamination and the islanders were removed in 1978. Presently, Bikini is inhabited by a handful of workers and dive-tourism staff but has not been officially resettled and cleanup plans have not been finalized (Guyer, 2001). Additionally, funding for ongoing health care and food supplement programs for dislocated islanders may be in jeopardy.[4]

Enewetak Atoll, RMI. Forty-two nuclear weapons tests were conducted at Enewetak Atoll from 1948 to 1958, including the 1952 detonation of the first full-scale thermonuclear device. Fallout contamination (fission products, activation products and unfissioned nuclear fuel) was highly localized because "large quantities of soil, water and lagoon sediment were incorporated into the ensuing fireball and fallout cloud" (Lawrence Livermore National Laboratory (LLNL), 2004). Enewetak was officially resettled in 1980 but questions remain about the adequacy of cleanup and the capability of the island to support the population it once did. The resettled population was monitored for internally deposited radionuclides from 1980-1997, by scientists from Brookhaven National Laboratory using whole body counting and plutonium urinalysis techniques (LLNL, 2004).

Maralinga Lands, Southwest Australia. The British nuclear weapon testing program in Australia and the Pacific consisted of approximately 21 major weapons tests and a number of smaller-scale but highly contaminating experiments conducted between 1956 and 1963. The Maralinga site, the only inhabited testing site in Australia, was intended to be a permanent atomic bomb testing site–the "Los Alamos of the Commonwealth." As a result, the Maralinga people were forbidden access to their traditional lands during the period 1952-1981. Cleanup campaigns were conducted during the 1960s and most lands were returned to the traditional owners, the Maralinga Tjarutja, in 1984; access remains restricted in 120 square kilometers of the test site's 3,200 square kilometers. Additional cleanup commenced in 1996 but ended in 1999 after a sub-surface explosion occurred during stabilization of debris pits at one of the sites; even after an extensive investigation the cause of the explosion was not determined and the cleanup of the remaining pits (according to the original cleanup plans) was never completed.

BACKGROUND ON WEAPONS TESTING

In 1945, the U.S. dropped two atomic bombs to end WWII, thus ushering the world into the Atomic Age. The destructive capacity of these bombs was awesome. However, the wartime use of the bomb raised questions within the military and the scientific community not about the bomb's force, but about its effectiveness, particularly on naval vessels. In January 1946, President Truman authorized plans for *Operation Crossroads,* the first nuclear weapons series in the Pacific and the second test since the 1945 *Trinity* test in Nevada.

The tests were initiated to increase the amount of "military and scientific knowledge of nuclear explosions" and "determine the effects of atomic weapons on naval vessels and the performance of military personnel" (Simon, 1997). Nevertheless, many felt that the origin of the tests did not stem from an empirical need to test the effects of the bomb on ships but rather "arose out of a bitter rivalry between the Navy and the Army Air Forces over the impact of the atomic bomb on their future budgets, missions and prestige" (Weisgall, 1994). Others felt that the tests were "a sign of post-war uneasiness, partially a result of advances in nuclear technology by our Soviet counterparts (Simon, 1997). Whatever the case, the tests were scheduled and approved. There was "no civilian agency within the government to question whether the tests made any sense" and no one to represent the interests of test site refugees (Weisgall, 1994). Authorization of *Operation Crossroads* was certified one year before the Atomic Energy Act would come into effect and transfer the control of nuclear weapons from the military to the civilian Atomic Energy Commission.

At the end of WWII, the U.S. was the only country with nuclear weapons capabilities. However, other countries were not far behind. The UK had partnered with the US in 1943 on the Manhattan Project and in 1947 authorized plans for nuclear weapons development. In 1948, the Soviet Union was the second country to successfully develop and test atomic weapons. The UK detonated its first atomic device in 1952. France followed in 1960, as did China in 1964 and India in 1974.

U.S. testing program in the Marshall Islands

In 1946, U.S. officials obtained the informal consent of Bikini leaders to temporarily relocate the entire population so that their lands could be used for nuclear testing,[5] with the promise of their eventual return to and the restoration of the islands to habitability. This consent was obtained only after Bikini was selected as the test site. In addition, the agreements concerning the relocation and use of the atoll were all verbal; the U.S. and Bikinians did not enter into a formal, written contract over possession and use of the atoll until 1956 (Kiste, 1968).

It is uncertain from historical accounts if the Bikinians truly had a choice of whether or not to leave the island; "it would not be implausible to suggest that they felt it would be to their disadvantage or perhaps beyond their power to resist" or stand up to a country that had emerged from the war as the most powerful nation in the world (Kiste, 1968). It is likely that the Bikinians felt that noncompliance with U.S. requests to relocate was beyond their power and in fact, would be to their disadvantage. U.S. government sources reported that the Bikinians "willingly agreed to move from their traditional homeland," understanding that "their sacrifice would contribute to world peace and be of benefit to all mankind" (Kiste, 1968).

When U.S. testing began in the Marshall Islands and the residents of certain atolls (Bikini, Enewetak and, later, Rongerik, Utrik, and Rongelap) were evacuated, relocation was viewed by all parties as temporary. Yet as the testing program progressed and expanded, the temporary nature of the relocation was reconsidered – by the U.S. In 1947 the U.S. was named trustee administrator of the Trust Territories of the Pacific (an area which comprised Micronesia in its entirety) and by 1948 U.S. officials had moved to make Enewetak Atoll a permanent proving ground.

U.S. officials had promised a temporary evacuation and a quick return of land to Marshall Islanders. However, the scope of testing changed and the "temporary" evacuation turned into a "relocation" that for some Enewetakese lasted thirty-three years.[6] The temporary evacuation of Bikinians has persisted for sixty years.

Much has changed on Bikini in those sixty years: three islands of the atoll were vaporized in one test, nearly 100 ships were sunk in the lagoon, all the vegetation was destroyed, and the topsoil was contaminated with the by-products of fission – cesium, uranium, strontium, and plutonium. In that time, the fabric of Bikini society has been rearranged and a traditional way of life is all but lost. Sixty years after the temporary evacuation and forty-eight years after the last test was conducted at Bikini, the Bikinians have yet to return.[7] The U.S. is bound by its initial promises but more specifically, by obligations spelled out in the 1986 Compact of Free Association (COFA), which calls on the U.S. to "...fulfill its responsibility for restoring Bikini Atoll to habitability."

The U.S. and Marshallese appear to have reached a stalemate about how to proceed with the cleanup, however. In 1998 the RMI, acting through the

Nuclear Claims Tribunal that was established under the COFA, adopted a strict 15 millirem/year effective dose equivalent standard for cleanup that the U.S. government has been unwilling to accept. Cleanup planning was soon put on hold. In 2000, the RMI filed a Changed Circumstances Petition with the U.S. government under terms of the COFA, an action that, combined with the hold in cleanup planning, shifts the focus toward additional claims for compensation for lingering health effects of radiation and away from negotiating the details of cleanup. For now, relations are strained over both U.S. rejection of the petition and the concept of "habitability" as it applies to the atolls most affected by the testing.

British testing program in Australia

In Australia, the situation with the British testing program is similar in some respects but different in others. In a test program that spanned eleven years (1952-1963), 3,200 square kilometers of desert in South Australia were contaminated with the by-products of testing – plutonium, americium, uranium, beryllium, and a host of other contaminants. The most serious and long-term damage (in terms of total land area and level of contamination) is the result of the series of "minor trials." These trials, which numbered about 200 between 1953 and 1962, were part of the UK research and development program designed to test weapons and weapons components. Although they did not involve nuclear detonations, plumes of plutonium dust[8] were spread over 500 square kilometers of land traditionally occupied by the Maralinga Aborigines (DEST, 2004). Among the worst of the contamination problems that exist today is that resulting from the deposition of some 22 kg. of plutonium in narrow surface plumes in the area known as Taranaki, the result of minor trials code-named Vixen B (Smith, 1991).

The Maralinga Aborigines – unlike the Bikinians – were not evacuated from their land specifically for the testing. Beginning in the 1920s, drought conditions forced Aboriginals off their land and all around Australia Aborigines were being moved from their traditional lands to mission outstations. The Aborigines of South Australia were moved to mission sites at Ooldea or fled to Port Augusta. The Maralinga Aborigines who moved first to the Ooldea mission were forced to move to a mission outstation at Yalata in 1952, when the UK government established the British Atomic Test Range and closed the Ooldea mission.

In the 1980s, there was a move on the part of the Aborigines in South Australia to leave the mission sites and return to a traditional lifestyle on their traditional lands. In 1984, the Maralinga people were granted 76,000 square kilometers of traditional land and in 1985 established a permanent settlement at nearby Oak Valley. Aboriginal control over the land was not absolute (Grabosky, 1989). Rather, access to land transferred to the Maralinga Tjarutja with passage of the 1984 Maralinga Tjarutja Land Rights Act was restricted due to contamination resulting from the British nuclear weapons testing.

On the Maralinga lands today, topsoil contamination is widespread and contaminated debris pits dot the landscape. The most intensely contaminated sites are the minor trials sites: Taranaki, Wewak, TM50, TM100, TM101, Kitten Lanes at Naya, and Kuli. Other sites are contaminated as well, but because these areas are remote and were never settled or used by the Aborigines, their cleanup has not been a priority.

SITE STATUS: CLEANUP AND RESETTLEMENT

Bikini Cleanup

Bikini's first cleanup was carried out in the late 1960's by the Officials of the Trust Territories, the U.S. Atomic Energy Commission (AEC), and agencies of the Department of Defense (DOD). The cleanup consisted of two overlapping phases (Kiste, 1974):

- *Phase one* (DOD and AEC): Removal of radioactive material and scrub vegetation and "in so far as possible," restoration of the islands' topography.
- *Phase two* (Officials of the Trust Territories, Marshallese Agricultural Department): Replanting of native vegetation, reconstruction of village infrastructure (concrete block houses with aluminum roofs), and relocation of the community.

The cost of the project exceeded three million dollars and took five years to complete. When a handful of Bikinians[9] resettled in December 1973 food crops and coconut had not yet had time to mature and bear fruit.

In vivo radiological monitoring of the population conducted in 1977 revealed a ten-fold increase in body burdens of cesium-137 since 1974. Tests conducted again in April 1978 revealed a further 75% increase and by September of the same year the Bikinians were evacuated to Kili. The source of cesium was determined to be then-mature coconut and food crops (Deines, Ann, David Goldman, Ruth Harris, & Laura Kells. (Preparers,

1990). An independent scientific consultant's report from 1982 indicated that radiation levels at Bikini exceeded U.S. radiation protection standards, results confirmed by Lawrence Livermore National Laboratory. Congress created the Bikini Atoll Rehabilitation Committee (BARC) in 1984 to investigate and report on "the feasibility and cost of rehabilitating Bikini Atoll" (Deines et al., 1990). BARC presented a suite of radiological cleanup options for Bikini Atoll in 1984 that were oriented toward reducing human exposure to contamination and/or eliminating contamination. The options included natural decay with delayed resettlement, leaching/salt-water irrigation, biological removal/cropping, topping, topsoil removal, chemical treatment, and a combination of partial topsoil removal and chemical treatment.

On principle, the Bikinians wanted –and still want— the U.S. to accept full remediation as a moral obligation. The 1986 COFA agreement included not only reparations payments, but in the Sec. 177 Subsidiary Agreement a call for eventual return of the displaced populations to their original places of habitation. As a result, many of the options presented in the BARC report (e.g., allowing the contamination to decay naturally over time or partial cleanup with restrictions on settlement and consumption of local food) were rejected. The Bikinians' first choice for cleanup involves removing 1.2 million cubic tons of contaminated topsoil from all of the atoll's islands;[10] this soil would be dumped at sea (or used to build a causeway between two of the islands) and replaced with 'clean' soil. The U.S. was – and still is – pushing a compromise involving removal of the topsoil from residential areas with chemical treatment[11] of topsoil elsewhere to block uptake of contaminants. The U.S. has argued that this compromise is logistically, technically, and scientifically sound and cost-effective, but it fails to address all of the Bikinians' concerns about contamination and risk. The 15 mrem/yr dose-equivalent standard that the Bikinians desire is a guideline that has not typically been applied to cleanups on the U.S. mainland. An equivalent standard was controversially applied at the Hanford Site in Washington State in an area that was cleaned up to a residential use standard (see Power chapter in this volume). A similarly strict standard was recommended for cleanup at the Rocky Flats, Colorado, site, but the standard ultimately selected was intended to protect a wildlife refuge worker and not a resident rancher (see Tuler and Webler, this volume). This choice was argued by the Department of Energy to be consistent with the intended use of the site as a wildlife reserve and not intended to support habitation in a manner that puts residents in more intimate contact with the land and its resources. The more the cleanup standard drops below the 100 mrem standard originally adopted, the more forgiving the environment becomes to cultural practices that might otherwise lead to harm. But cleanup costs may also rise dramatically (Burger et al., this volume).

U.S. officials point out that removal and disposal of all the topsoil would not only be cost-prohibitive but would also result in major environmental damage that would require decades of rehabilitation–longer than it would take for the most biologically active remaining contamination to decay naturally. It also creates a new problem – what to do with 1.2 million cubic tons of contaminated topsoil.

Topsoil removal was rejected as an option for cleanup of the UK test site at Marlinga:

> Recognizing the wider goals of protecting the environment as well as people, there is no point in solving one problem by creating another. Land remediation can do far more damage to soil and plant communities, which are valued by traditional land owners, than the residual radiation ever will (Daniels, 1997).

At present, Bikini Atoll's main islands, Bikini Island and Eneu Island, *are* technically habitable, even if judged by the U.S. EPA's recommended 15 millirem limit for radiation exposure. The basis for this conclusion however is a series of risk scenarios that assume the restrictions on the consumption of local food (coconuts in particular) and locally drawn water noted above. In fact, Bikini now has less background radiation than some American cities (Kristof, 1997). The problem, however, is that surface soils still contain cesium-137. Given the history and the cultural importance attached to the consumption of locally grown foods, especially coconuts, islanders are reluctant to return. The failed 1973 relocation attempt is still a vivid memory for many Bikinians. The Marshallese are disinclined to accept even the most expert advice or enter into agreements concerning another relocation attempt.

They are also reluctant to take actions that might limit or terminate the US role in maintaining health care, food, agriculture and water programs that have been set up in the time since the first relocation.[12] To the contrary, Bikinians are anxious to extend the United States' formal obligation to restore the atoll and ensure a long-term commitment to 'caring for the Bikinians' as they promised in 1946 when US officials told them: "Never mind if you are living on a sandbar or even adrift on a raft at seas, we will take care of you as if you were our very own children" (Weisgall, 1994).

Without treatment the local food sources that are contaminated will remain so for decades;[13] settlers of Bikini would still have to rely entirely on imported food and water to meet radiation exposure limits. The Nuclear

Claims Tribunal's[14] adopted 15 millirem radiation protection standard is in line with a U.S. EPA guideline for radiation exposure protection, , but the original resettlement trust fund figures were calculated in 1988 for a less restrictive 100 millirem radiological cleanup standard.

Complicating matters further, the funds originally provided by the U.S. under Sec. 177 of the COFA, totaling $150 M, were to include compensation for health effects from exposure to contamination. But they were set largely without estimation of the actual number and nature of health claims that might result. The intent was simply to compensate for personal injury on the basis of a pre-set schedule that was part of the agreement. The Claims Tribunal soon exhausted the available funds. Most recently, in settlement of class actions brought before it, the Tribunal awarded $91 M to the people of Enewetak (in 2000) and $251 M to the people of Bikini (in 2001). These amounts included costs for cleanup and restoration at both former test sites and were well in excess of the funds available and thus could not actually be paid. In rejecting the RMI's Changed Circumstances Petition, the Bush Administration claimed that the U.S had spent $531 M on compensation and other assistance in the Marshall Islands since 1958. The petitioners on the other hand estimated that the total unpaid obligations were well in excess of $1B (Congressional Research Service, 2005).

It is uncertain how many of those who identify themselves as Bikinian[15] would come back to the atoll were it open to unrestricted settlement.

> Everybody will go back, some for a short visit, others for a long time. But you won't see 2,000 people jump on a big ship and sail home to live. It's not conceivable that they'll go back to living the way they were. They're part of the modern world now (Bartholet, 1997).

There is a contingent of community elders, many of whom were themselves evacuated in 1946, who are the true driving force behind rehabilitation efforts and many of them are at the end of their lives. Regardless of how many Bikinians want to and are able to return to live on Bikini, there is a strong sentiment among Bikinians that on principle they should have the option to return. Despite these concerns for contamination, Bikini today does not lie uninhabited; it has become a resort center for international dive tourism, an industry more able to accommodate the current restrictions on the consumption of locally grown food and water. A few Bikinians are able to live on the island, but as employees of the tourism broker, with food and lodging provided by tourism industry employers. The paradox is that "the same atomic bombs and hydrogen bombs that once caused immense tragedy to the Bikinians…have now made the island an unusually valuable tourist property" (Kristof, 1997).

Enewetak Cleanup

Enewetak Atoll, used for testing in the latter part of the program, was cleaned up according to international radiation protection standards and the U.S. Department of Energy criteria in the late 1970s. Its inhabitants were able to resettle in the 1980s, although inadequate food supplies forced 100 individuals to leave Enewetak in 1981.

Enewetak's cleanup, planned in 1972 and executed in 1977–1980, consisted of isolating topsoil from five islands, collecting contaminated debris (concrete, metal debris, cables, bunkers, buildings, etc) and depositing it all in a crater carved out of Runit Island by an 18 kiloton detonation called the Cactus shot, one of the many test shots of the *Hardtack I* test series. For those that have resettled Enewetak the main concern is the integrity of this capped crater, which contains 40,000 cubic meters of contaminated soil and debris.

In theory, the crater contains the most severely contaminated debris. However, during an earlier test (the *Quince* shot, 1958), large amounts of plutonium nuclear fuel were scattered over Runit Island when a nuclear device only partially detonated. Twelve days later, in preparation for another test, the top 8-15 inches of highly contaminated soil on Runit were bulldozed and dumped into the lagoon, thus making the lagoon the largest reservoir of plutonium contaminated material on the atoll today.

Core samples taken of the capped debris mound on Runit (called the Runit Dome) in 1990 revealed that while the dome has maintained its structural integrity, the stabilization techniques did not work as expected. Following the "Tremie Method", the crater was filled with soil and debris mixed with water and concrete. This slurry was then covered with a layer of soil and concrete; a layer of water and miscellaneous larger debris went on top of that. A hole was left in the middle of this mound that was filled with more soil and debris before the structure was capped and sealed with 18-inch thick concrete panels. The layers were supposed to cure and solidify, but the 1990 sampling revealed otherwise. In addition, data from the core sample indicated that the amount of radioactivity contained within the crater and dome was roughly equivalent to the environment *outside* the crater.

This engineered barrier has nonetheless been judged effective in preventing the release of contamination from the crater and dome. Scientists maintain that the dome presents no risk.

Any fear that this structure contains amounts of radioactivity whose release would cause damage to the environment that would result in a greater effect on human health is unfounded (Noshkin & Robison, 1997).

Residents of Enewetak are concerned, however, that Runit Dome is deteriorating and feel that an aging structure filled with over 40,000 cubic meters of debris represents a hazard. They have expressed dissatisfaction at what they perceive to be a lack of consistent monitoring of the site: "The people are convinced that it must be one of the most dangerous places at the Atoll since the United States spent millions of dollars to contain the radioactive material..." (Noshkin & Robison, 1997). Regular surveillance and monitoring took place for four years after the structure was erected; monitoring has continued only intermittently with no indication of significant release or potential for release. Despite assurances from scientists that exposure limits are within safe parameters, the Enewetakese remain uneasy about their resettlement.

Maralinga Cleanup

Site cleanups and surveys of the British Atomic Test Range at Maralinga were conducted on the contaminated land in 1964, 1966 and 1967.[16] The Australian Weapons Test Safety Committee (AWTSC) recommended acceptance of the 1967 cleanup as the final "definitive" cleanup; fences and signs identifying contaminated areas were to be removed. However, the Australian Ionising Radiation Advisory Council (AIRAC), AWTSC's successor, favored "maintenance of fences around radioactive disposal pits and warning signs...at ground zeroes and burial sites" (Davy et al in Mohr et al., 1997). Concern for the final state of the site did not go away however, and a new round of cleanup was initiated in the 1990s, but abruptly terminated due to unforeseen technical problems. Fencing off of the most contaminated areas and long-term management will still be required if the Maralinga people are to resume traditional practices on these lands.

Operation Hercules (1964) had the "single objective of removing the major hazards so that entry to the area, within the sense of military operations, could be made without direct Health Physics supervision" (Mohr et al., 1997). Debris was moved and structures were stabilized, but not much else was done in the way of addressing residual contamination of the vegetation and soil in the test area. A site survey, *Operation Radsur*, was conducted as a follow-up in 1966. *Radsur*'s goal was to better define "the prevailing levels of exposure rate and contamination" (Mohr et al., 1997).

A more thorough cleanup, *Operation Brumby*, was undertaken in 1967. However, subsequent investigations into this operation indicated that a

> discrete, potentially recoverable, quantity of plutonium remained at the range....accessible to terrorist groups...and held in a manner inconsistent with Australia's obligations to the International Atomic Energy Agency (IAEA) as a signatory of the Non-Proliferation Treaty (Davy et al. in Mohr et al., 1997).

This plutonium was recovered and repatriated to Britain in 1979 but the interest of the Maralinga people in the cleanup – and the contamination – remained.

Following the 30th anniversary commemoration of the first detonation of the UK testing program in Australia (1983), the Maralinga people were granted their land back and Maralinga representatives were sitting on the *Royal Commission into British Nuclear Tests in Australia*. The Maralinga Commission (of the Royal Commission) consisted of representatives from traditional owners, the UK government and the Australian government. Because the contamination extended onto traditional lands, the Maralingans' interests concerned not only direct exposure as a result of the test program's first detonation (*Totem*) which blanketed the desert with a "Black Mist", but also the contamination of their traditional lands at the minor trials sites (Davy et al. in Mohr et al., 1997).

In 1984, AIRAC built an extensive but temporary fence to enclose contaminated areas at the Tarakaki site. More permanent fences were erected in 1987 that enclosed 21 debris pits, firing pads, and other highly contaminated areas. The Maralinga Rehabilitation project was undertaken in 1996 with the goal of completing the cleanup process and opening the area to habitation, or "to reduce the risk arising from radiation exposure of individual Aborigines, living an outstation lifestyle, to a level that was acceptable to the Aboriginal Community and the Australian Government" (ARPANSA, 2004). As a recent commentator noted however, "the boundary markers might have a life of 50 years, but half of the plutonium will still be there in 24,000 years" (Parkinson, 2002).

When the Australian government established the Maralinga Rehabilitation Technical Advisory Committee (MARTAC) in 1993, it also convened the Consultative Group–whose membership included representatives of the Maralinga Tjarutja Aborigine people–that had also worked on cleanup with the government's Technical Assessment Group (TAG) during the 1980s. The cleanup goal established by the TAG was to reduce risk to levels acceptable to both the government and the Aboriginal community, for individuals living an outstation lifestyle (TAG, 1990). The rehabilitation plan, as it was originally envisioned, involved excavation and consolidation of debris pits, removal of highly contaminated topsoil from certain areas, re-burial of soil and debris, and stabilization of contaminated materials at the Taranaki site through the use of in-situ vitrification (ISV) technology.[17] The plan allowed for cleanup of most sites with the goal of restricting access to only the most contaminated sites (DEST, 2004; Minchin, 2000).

The project was going well until a subsurface explosion occurred as the eleventh pit was being treated at Taranaki. The equipment suffered significant damage but no injuries were reported. Cleanup was suspended and an extensive investigation was initiated. The investigation did not come up with any conclusive evidence concerning the source of the explosion and the Commonwealth government decided to abandon stabilization treatment. The remaining excavated debris at the Taranaki site was simply re-buried without further treatment (DEST, 2004).

In all, over 350,000 cubic meters of contaminated surface soil and debris were removed from the test areas and buried in trenches 10-15 meters deep and capped with 5 meters of clean soil. Eleven of the 21 trenches were stabilized using the in-situ vitrification process; markers and signs were erected to mark the location of each trench; and an extensive revegetation project was carried out. Boundary markers "discouraging permanent habitation" were also erected (DEST, 2004).

In-situ vitrification had been selected as a stabilization treatment because it was thought that the pits were highly contaminated with plutonium; it was also believed that the process could be carried out without placing workers at risk. The 1999 explosion proved otherwise and subsequent research revealed significantly lower levels of plutonium contamination than was originally thought.[18]

Despite the termination of the stabilization treatment, the Maralinga Rehabilitation Project is finished; on the books it came in on schedule and slightly under budget.[19] The Commonwealth Government agreed to a $A13.5 million payment over two years to compensate for loss of access to the restricted area and to fund outstations, improve health services and water supply systems, and develop infrastructure. In addition, the Maralinga Tjarutja accepted a 1993 *exgratia* payment of 20 million pounds towards cleanup (Minchin, 2000; O'Conner, 1997). The Maralinga Tjarutja has unrestricted access to 90% of formerly restricted lands as a result of the cleanup program. Occupation is restricted for 120 square kilometers at the Taranaki site (DEST, 2004; Minchin, 2000).

The Australian Radiation Protection and Nuclear Safety Agency (ARPANSA) calculated radiation doses – effective dose values for inhalation pathways – for a range of sites at Maralinga following the cleanup. Because it was impossible to determine factors influencing future occupancy of Maralinga areas by Aboriginal communities, the agency assumed a full occupancy rate and considered inhalation to be the most likely risk pathway.

The agency reported that pre-remediation dose estimates revealed that certain areas (Taranaki, Wewak, TM100 and TM101) were found to have inhalation dose rates that were unacceptably high under all but the most rigorously controlled circumstances. Following the rehabilitation, all areas at Maralinga were shown by the dose assessment to be "well within acceptable limits for all envisioned land uses" (ARPANSA, 2004). In 2000, the Australian government declared the cleanup to be complete and all but 10% of the Maralingas' former land to be fit for habitation. Current restrictions on permanent occupancy, the government attests, can be seen as "purely precautionary measures as doses due to inhalation for permanent occupancy of all but a few areas....are all well below the 1 mSv/y limit for members of the public"(ARPANSA, 2004). The Australian government holds that the cleanup— consisting of vitrification at 11 sites and exhumation of plutonium and reburial in trenches at the remaining sites- was carried out according the best scientific advice and is consistent with 1992 code of practice.

Not all are happy with the results of the rehabilitation project, nor do they agree with the government's assessments. Four of the five members of the government's technical advisory committee that originally drafted the code say that it was solely intended for the near surface burial of short-lived radioactive waste and was "never meant to apply to the burial of plutonium-239" (Pockley, 2000). Furthermore, the "calculated dose of an Aboriginal living a semi-traditional lifestyle...is five times that allowed for a member of the public" (Parkinson, 2002). In regard to land restriction, the area within the boundary markers which remains contaminated at above-acceptable dosage is much smaller than the enclosed area to which access is restricted (ARPANSA, 2004).

Nevertheless, at this point all parties seem willing to move forward under the presumption that the task at hand is now to determine the basis for living with the contamination that remains and not to plan additional cleanup. MARTAC, the Maralinga Rehabilitation Technical Advisory Committee, held its final meeting in 2001. A massive report issued by the Commonwealth Government of Australia in 2003 (MARTAC, 2003) was published for the express purpose of compiling for posterity the key documentary record on site remediation:

> Poor British record keeping impeded MARTAC from its first meeting to its last. To avoid a similar situation in the future, MARTAC therefore decided to bring together ... all the final reports that were prepared by various contractors to detail the work they preformed for the Maralinga rehabilitation program" (MARTAC 2003, p. vi).

A WIDER HISTORICAL PERSPECTIVE

This recent history – from the beginning of the testing programs to the present day – is well documented (Deines et al., 1990; MARTAC, 2003; Mohr et al., 1997; Neidenthal, 2001; RMI Online, 1997; Simon, 1997). But looking at the situation in the context of the past 50-60 years tells only part of a larger story. Land in the Pacific Islands and in Australia is critically important to culture and cultural history. It is the currency of cultural identity, social status and political hierarchy. Oral histories and interviews, taken by a number of researchers, anthropologists, and officials over the years indicate that to the former residents, Bikini Atoll, Enewetak Atoll and the Victoria Desert are more than just plots of land. To them, land is:

- "…an extremely limited resource…'the source of their existence…their most valuable asset" (Kiste, 1968).
- "…a treasure from our ancestors, handed down over many generations from the beginning of time" (Dretin Jokdur, the *iroij* (traditional leader) of the Bikini people, December 12, 1997, in Neidenthal, 2001).
- "…like a relative to us: like a father or a mother or a sister or a brother, perhaps more like a child conceived from our own flesh and blood" (Bikinian Lore Kessibuki, 1997 interview in Neidenthal, 2001).
- "…the place God has given to us" (Bikinian Pero Joel, 1989 interview in Neidenthal, 2001).
- "…gold – if you were an owner of land you would be held up as a very important figure in our society. Without land you would be viewed as a person of no consequence" (Bikinian Jukwa Jakeo 1987 interview in Neidenthal, 2001).
- "…central to Aboriginal people. It is the basis of all cultural, economic, social and political activity and many people living on Aboriginal lands" (O'Conner, 1997).

The importance of a return to traditional lands and a return to a traditional lifestyle for the Bikini, Enewetak, and Maralinga people is connected to a long and complicated history. Cultural identity amongst Pacific Islanders and Aboriginal Australians is a geographical identity that flows from memories and values attached to places. Oral traditions, religious and spiritual beliefs, superstitions and lore often trace the evolution of cultural identity through sacred symbols and sacred places.

While the dates and events discussed previously are revealing, they tell nothing of how the people of Enewetak or Bikini came to live on their respective island, how land was passed down through generations, how the traditional ruling system dictated land ownership and territoriality, or how losing one's land can mean losing control of one's destiny.

In coastal villages and island nations the sea is a guide to social history. Certain reefs, channels, islands, passages and seamounts are associated with particular spirit beings. Visiting and revisiting these sacred places plays a crucial role in mediating history and in the formation and maintenance of cultural identity. The assumption that "the movement of a small population was a relatively simple matter which could be accomplished swiftly and with little planning or difficulty" has been proven horribly wrong (Kiste, 1974). Pacific Islanders have maintained a close symbolic and emotional attachment to the land and the surrounding seas (Feinberg, 1995).

> Everywhere there is a spiritual attachment to the land and the surrounding seas unparalleled in Western or other societies...loss of land and its resources can be culturally and political devastating to island societies (Dorrance, 1992).

Islanders rely heavily on sparse land resources and inshore fisheries for subsistence; it is their food, their shelter, and their transportation.

In the case of these former nuclear weapons test sites, returning the land to the traditional owners has been a high priority for site rehabilitation and remediation – a priority that has proven particularly challenging. Much of the focus in the Marshall Islands has been on technical feasibility and cost, while relatively little has been done in the way of collecting ethnographic data or getting input from the traditional landowners. Cultural barriers have presented unique problems and have resulted both in remediation setbacks and resettlement failures.

The Maralinga case exemplifies how stakeholder participation can be used to overcome cultural barriers, despite technical obstacles. Cultural sensitivity training, ethnographic research, and the involvement of the Maralinga Aborigines in the decision-making process has resulted in a new agreement between the Commonwealth, the South Australia Government, and the Maralinga Tjarutja, which permits unrestricted access in 90% of the former restricted area.

In each of the cases, a fundamental mistrust of official solutions on the part of traditional landowners has been evident. A poor record of success, ill-managed risk strategies, and lack of stakeholder participation and consultation have frequently contributed to this mistrust. LaPorte and Metlay (1996) identify several reasons that relationships between the public and the decision-makers breakdown. They point to situations where:

> There is a mismatch in the distribution of benefits and costs (in financial and social terms), [and when] the risk or hazard associated with significant program failure appear high or long-lasting (LaPorte & Metlay, 1996).

This appears to be the case for Bikini Atoll in particular. Amata Kabua, former President of the Congress of Micronesia and President of the Marshall Islands (1979-1996) commented on this in 1968:

> ...it has come to the point where it is difficult for the Micronesian to believe anything the U.S. administration says—everything smacks of idle promises we have been listening to for the past twenty years. The result of this is that the Micronesian no longer takes the U.S. administration seriously (Kabua in Dorrance, 1992).

The failed 1970s relocation attempt at Bikini underscores this point and further feeds feelings of mistrust. Because what was believed to be expert scientific advice failed to protect returning landowners in the past and because the problems associated with contamination and risk are highly technical in nature, the Marshallese are reluctant to accept any "final" cleanup settlement that is not based on the most rigorous safety standards that have been applied on the U.S. mainland.

Overcoming this reluctance and the obstacle of mistrust depends on mutual respect and understanding – something that was given little attention in early efforts. Attempts by U.S. officials to introduce complex concepts of risk, risk management, and dose response analysis resulted in confusion, frustration and negative feelings on the part of the Bikinians.

> Government officials had expected the resettled Bikinians to be aware of the risks of returning to their island home, to self-police their food consumption habits (despite little effort at risk communication) and to accept the risks of radiation exposure in exchange for the benefits of being able to return to their ancestral home. These assumptions proved to be in error (National Research Council, 2000).[20]

Bikini is judged to be safe for human habitation today, with the proviso that food and water be imported and consumption of locally grown food and locally caught fish be restricted. Bikinians remain reluctant both to return to the Atoll and accept the conditional restrictions on their return.

> For people with very limited understanding of radioactivity and radiation physics, and no words to express such concepts in their own language, trust in the messenger may be more important than the message for its acceptance (McEwan et al., 1997).

For the Australians, the situation in many respects was no different.

> One of the 'dread' factors regarding radiation hazards is the perception that it is ultra-technical; for example, it has its own language...this reinforces the view that radiation hazards are something special and scary (Daniels, 1997).

To their credit, Australian and British officials recognized early on the importance of participation on the part of the Maralinga Aborigines in the decision-making process.

> A most positive aspect of the process was the political commitment to the inclusion of the Aboriginal interests...this commitment was matched at the senior scientific level (Davy et al., 1997).

Complex concepts of risk and radiation hazards were translated into terms the Aborigines could understand and incorporate into their daily lives.

> Once the discussion came down to what the traditional owners could or could not do in [a particular] area, the discussions proceeded in a very detailed fashion (Daniels, 1997).

The situation at Maralinga seems equally serious in terms of the contamination yet conflicts over resettlement and future land use have been fewer in comparison to those in the Pacific Trust Territories. And while the process of carrying out the cleanup in Australia was marked with a certain level of controversy and remains the subject of speculation and internal investigation, the Maralinga Aborigines seem satisfied with the decisions made about their land and many have returned to a semi-traditional lifestyle.

Stakeholder involvement in decision-making was essential, particularly with regard to assumptions of future land use. Then there was the question of the technical and scientific approach. Given the failures of the past, it was critical that the approach should have social and political defensibility; it needed to be accountable, transparent, and credible (Daniels, 1997).

Now the arena has shifted to detailed planning for site use and concurrent environmental management consistent with the residual contamination at the site. The Commonwealth and South Australalain Governments and the Maralinga Tjarutja have jointly drafted the *Maralinga Land and Environment Management Plan*, intended to serve as "the governing plan for post-rehabilitation management of the resources at Maralinga and Emu" (MARTAC, 2003). In broad structure at least, this plan very much reflects the planning template recommended by the National Research Council in its long-term institutional management study (NRC, 2000).

SUMMARY AND CONCLUSION

Marshall Islanders and Australian Aboriginals that were relocated or otherwise deprived of access to traditional lands during the nuclear testing programs have long histories and significant attachments to their lands that have perhaps not been fully considered in the development and implementation of remediation and resettlement plans. The incorporation of lifestyle considerations and ethnographic data for the specific population must be considered in the development of risk analysis and determination of exposure pathways.

The failure of the initial resettlement at Bikini Atoll was a result of not incorporating accurate dietary information into the risk analysis that was done, not recognizing the different behavior of contaminants in calcium carbonate-rich atoll soils and not consulting closely with the Marshallese. At Maralinga, the initial path was very similar, but with the passage of time the problem came to be seen in a different way with the result that key decisions were made in close consultation with the Maralinga Tjarutja. A more mature approach to risk assessment than used initially by either British or American authorities was adopted, with ethnographic data utilized in examination of exposure pathways.

As the cases examined in this paper illustrate, long-term stewardship must be approached from a culturally and historically informed perspective. Narrow conceptions of LTS that fail to focus on effective communication and do not allow for stakeholder participation run the risk of unacceptable outcomes. At Bikini and Enewetak, a lack of trust contributed to a failure to choose a final resettlement strategy.

We present the case that to design a successful remediation and resettlement program the agencies involved must consider:

- The value of the resources they are proposing to rehabilitate or replace, in the cultural context of displaced inhabitants;
- How testing and the presence of the government agencies has changed the social fabric of the population in question as well as the landscape;
- The relationship of traditional landowners to that landscape;
- The need for communicating accurate information about risk; and
- The need for believable assurances that the lands can be resettled safely, coupled with recognition that believability is highly dependent on the level of trust that exists.

The challenge then becomes identifying a solution that meets not only the technical and scientific requirements of national and international law but also addresses the social and cultural needs of traditional landowners.

For Enewetak and Bikini the cultural and social needs of the evacuees and their descendents have been addressed through food and health programs, which have provided medical care and funding for supplemental food importation, monetary compensation and environmental monitoring. What has not been given and what the Bikinians specifically are demanding is explicit assurance from the U.S. government that the land is safe for habitation, specific assurance for ongoing funding for health and food programs, and explicit recognition of the sacrifices that the Marshallese have made for the United States. Questions of reparations are intimately bound up with questions of remediation and restoration.

In the Maralinga case, the value of ethnographic research, cross-cultural communication and stakeholder participation was recognized early on and the Maralinga people were eventually involved in planning the cleanup and resettlement strategy to productive effect. Such involvement and communication led to understanding of native connections to the land and their traditional practices, thus enhancing the likelihood of resettlement success. Indeed, only at Maralinga has the population returned to a semi-traditional lifestyle on a majority of their former lands.

Yet this 'success' too is more nuanced than it may appear. It brings to light conflicting definitions of habitability that have led to an outcome that likely still compromises the value of the indigenous culture and traditional lifestyles to a degree. The Australian government was effective in garnering an understanding of the Maralinga people's cultural practices and how these practices may affect long-term use of the site. The fact that cultural practices may put the population at risk was understood; in fact, the population was warned that traditional practices at contaminated sites could increase risk. In restricted areas, barriers were put in place warning of radiation hazards "above those considered safe for permanent habitation" (Thiess Services, 2003). Meanwhile,

> It is now impossible for casual visitors making intermittent forays into the area, for example tourists, geological prospectors and surveyors, who do not engage in abnormal dust raising or large-scale soil-disturbance activities to receive a committed dose by inhalation" of the 1mSv/y limit (ARPANSA, 2004).

Essentially, the Maralinga were presented with a choice between culture and risk: they could return to their former lands, but not fully to their traditional lifestyles. In reference to the factors that may affect exposure risk in coming years, the Australian government acknowledged that the degree to which exposure pathways contribute to the potential dose depends on the "type of lifestyle practiced by occupants of the land" (ARPANSA, 2004).

> Lifestyle changes could also markedly affect dose estimates. If in time the Maralinga Tjarutja were to move towards a more European lifestyle, with extensive areas being covered by concrete, tarmac, buildings and lawns, and living in western-style houses in suburban settings, then the dust levels and hence doses are expected to be much lower (ARPANSA, 2004).

Hence the conception of habitability that guided the Maralinga Rehabilitation Project and led the government to hail the project's completion an "unparalleled success" may still raise fundamental questions regarding cultural value. Resuming traditional lifestyles can only be done at the assumption of greater risk. The presumption is that this trade-off has been duly considered and accepted through the extensive involvement that members of the Maralinga Tjarutja had in decisions about the cleanup and rehabilitation of the site.

In the case of the Bikinians, return under present circumstances would mean at a minimum relinquishing the tradition of coconut eating that has been central to their cultural heritage, and accepting a high degree of dependency on the outside world for essential foodstuffs. Return to the island seemingly entails a choice: to accept the higher risk of eating contaminated fruits and other practices that preserve traditions, or to give up these practices and assimilate to the norms of the modern world. Especially notable in the future of Bikini is international dive tourism—through a commercial enterprise that since 2000 has been owned by the Bikini Atoll Local Government Council and which is bringing benefits in the form of revenues and employment for young, educated Bikinians who might otherwise leave for opportunities elsewhere (http://www.bikiniatoll.com, accessed May 7, 2006).

This risk to cultural identity, and the choice between risk and tradition, is a common thread that runs through each of the cases in this analysis and raises issues central to long-term stewardship of contaminated sites. Understanding the significance of land and land ownership to dislocated populations, as well as the social impact of years of exile from homelands, is essential to understanding the depth and breadth of the situation today. *What next? How clean is clean enough? Will the resettlement be sustainable? Will the resettlement preserve cultural identity and value?* These questions can only be answered in social and cultural as well as biophysical context.

NOTES

1. In this paper the term "traditional owners" refers to Bikini Islanders, Enewetakese and Australian Aborigines who have occupied lands in the Bikini Islands and Enewetak Atoll (in the Marshall Islands) and the Maralinga Tjarutja of Southwest Australia, respectively, for hundreds of years whether or not they "owned" the land in the contemporary understanding of ownership.

2. Standards for the protection of human health have been promulgated by various agencies: United Nations Scientific Committee on the Effects of Atomic Radiation (UNSCEAR), International Atomic Energy Agency (IAEA), International Commission for Radiation Protection (ICRP), and the Nuclear Regulatory Commission. IAEA recommendations are outlined in the 1996 publication International Basic Safety Standards for the Protection Against Ionizing Radiation and for the Safety of Radiation Sources, Safety Series 155, and the Australian National

Health and Medical Research has published standards that were used at the Maralinga site, the Code of Practice for the near-surface disposal of radioactive waste in Australia.

3. Bikinians were moved to Rongerik, Kwajalein, Kili, back to Bikini, and to Kili again. A handful of Bikinians moved back to Bikini and lives and works there today. The Enewetakese were moved to Meck Island (Kwajalein), Ujelang, back to Enewetak (1980) and to Ujelang again in 1981. Ujelang Islanders were moved from their home to make room for the Enewetakese and again for a nearby weapons test. Rongelap Islanders were moved to Kwajalein and Ailliangenae, back to Rongelap in 1957 and back to Kwajalein in 1985.

4. The U.S. and the Marshall Islands have been bound by provisions set out in the Compact of Free Association (COFA 1986–2001) which was amended and reinstated in 2004. The Compact is a bilateral agreement through which the U.S. provides a defense commitment and substantial economic assistance to the Republic of the Marshall Islands. The amendments leave some remaining questions about the source of funding for health care programs and food supplement programs for residents of the four main atolls affected by nuclear testing. Section 177 of the original compact outlined a program whereby the U.S. would distribute $30 million over 15 years ($2 million a year, from 1986–2001) to assist with health care programs and "services related to the consequences of nuclear testing in its health care system" (Public Law 99-239 Joint Resolution to approve the "Compact of Free Association"). The amended COFA includes funding for a Marshall Islands health care services program but does not include funding earmarked specifically for the population covered under the original '4 Atoll/Section 177' health care program and allows for environmental monitoring at only the most basic levels.

5. The population living on Bikini Atoll in 1946 consisted of 167 individuals from eleven families. The entire population, including those living away from the atoll at the time, was 196 individuals.

6. Enewetak was resettled in 1980 but 100 individuals returned to Ujelang in 1981 because of inadequate food supplies.

7. "Of the original 167 residents 70 are still alive, and the population has expanded to 2800. Most now live in Kili and Majuro; some live in the United States" (Guyer, 2001).

8. The majority of the plutonium contamination is a result of plutonium dispersal that occurred in about 20 of the 200 experiments conducted in the Minor Trials (DEST, 2004). Most of the contamination is Plutonium 239, which has a 24,000 year half-life (Minchin, 2000).

9. The Bikini population had grown to over 400 individuals. 141 returned to their traditional home in 1973; the rest remained on Kili, refusing to return "until they received appropriate compensation for the past 25 years" (Deines et al., 1990).

10. This plan is sometimes referred to as "The Big Scrape."

11. Treatment proposed in this scenario is potassium fertilization. Because atoll soils are potassium-poor and because potassium is structurally similar to cesium-137, plants (especially coconut plants) on Bikini are taking up cesium instead of potassium. That cesium is concentrating in the plants' fruit. The cesium is taken up by plants because it is readily available, is water soluble, and is only weakly bound to atoll soils. "In a potassium-deficient media, cesium acts as a replacement for potassium, but only

to a limited degree....When both elements are available, plants selectively absorb potassium and discriminate against cesium" (Robison and Stone 1992).

12. United States PL 96-205, enacted March 12, 1980, provides that: " In addition to any other payments or benefits provided by law to compensate the inhabitants of the atolls of Bikini, Enewetak, Rongelap, and Utirik, in the Marshall Islands, for radiation exposure or other losses sustained by them as a result of the United States nuclear weapons testing program at or near their atolls during the period 1946 to 1958, the Secretary of the Interior shall provide for the people of the atolls of Bikini, Enewetak, Rongelap, and Utirik and for the people of such other atolls as may be found to be or to have been exposed to radiation from the nuclear weapons testing program, a program of medical care and treatment and environmental research and monitoring for any injury, illness, or condition which may be the result directly or indirectly of such nuclear weapons testing program."

13. Natural decay of radioactive contaminants at Bikini to levels deemed safe under international standards is expected by 2064.

14. The Marshall Islands Nuclear Claims Tribunal was established in 1988 with jurisdiction to "render final determination upon all claims past, present and future, of the Government, citizens and nationals of the Marshall Islands which are based on, arise out of, or are in any way related to the Nuclear Testing Program." The Tribunal is authorized by statute to issue regulations "establishing a list of medical conditions which are irrefutably presumed to be the result of the Nuclear Testing Program" (Nuclear Claims Tribunal, 2004).

15. The Bikini population in 2003 was reported to be 3300. As of November 2003 the Bikinian demographics are as follows: Kili Island 1,175; Majuro Atoll 1100; Ejit Island (of Majuro Atoll) 275; other Islands in the Marshalls 275; U.S. and abroad 475. Total Bikinians in 1946, 196; Bikinians living on Bikini in 1946, 167. Bikinians living that were removed from the island in 1946, 63 (104 deceased) (Neidenthal 2001).

16. Operation HERCULES, Operation RADSUR, and Operation BRUMBY, respectively.

17. In-situ vitrification is a process that involves passing an electric current through electrodes in the ground to melt soil and debris and incorporate the material into a vitrified monolith (a glass/ceramic block), thereby immobilizing the radiological contaminants (DEST, 2004).

18. "Originally, it was thought that about 20 kg of plutonium would be concentrated at the Taranaki site, much of it buried in the Taranaki pits. In fact, only an estimated few kilograms of plutonium has been recovered during the remediation including: about 2.6 kg of plutonium buried in the soil disposal trench at Taranaki [and] about 0.6 kg of plutonium buried in the debris disposal trench (0.4 kg of which is in vitrified blocks)" (Minchin, 2000).

19. The Maralinga Rehabilitation Project budget for 1996-1999 was $A108 million (in 1999 dollars) (DEST, 2004, Minchin, 2000).

20. Ironically, the conditions created by testing and the conditions imposed by the need for continued radiological protection have proved well suited to a newer group of visitors to the remote atoll; Bikini has become an international tourist destination for underwater wreck diving (Jennings and Leschine 2000, Kristof, 1997, Perrine 2001).

ACKNOWLEDGMENTS

This chapter is derived from an unpublished paper originally prepared for the Consortium for Risk Evaluation with Stakeholder Participation (CRESP) Public Policy Center, University of Washington, Seattle with funding from the U.S. Department of Energy. The authors are grateful to Philip A. Okney, Esq. of the Republic of the Marshall Islands Nuclear Claims Tribunal for helpful comments on an earlier draft of this chapter. The opinions expressed are entirely those of the authors, and not the Department of Energy, the Claims Tribunal, or Mr. Okney.

REFERENCES

ARPANSA. (2004). *Maralinga Site Cleanup: Fact Sheet*. Australian Radiation Protection and Nuclear Safety Agency, Commonwealth Department of Health and Aged Care. Available at http://www.arpansa.gov.au/pubs/basics/maralinga.pdf (accessed December 2004).

Bartholet, J. (1997). "Bikini Atoll's Blue Lagoon." Newsweek (International Edition). December 12.

Buell, L. (1998). Toxic Discourse. *Critical Inquiry, 24*(3), 639–665.

Congressional Research Service (CRS). (2005). Republic of the Marshall Islands *changed circumstances petition* to the Congress. CRS report for Congress Order Code RL 32811. Congressional Research Service, the Library of Congress. Washington, D.C.

Daniels, J. (1997). Summary panel discussion. In: L. C. Mohr, W. Nixon, & D. Hoel (Eds), *International case studies in risk assessment and management*. Charleston, SC: The Medical University of South Carolina Press.

Davy, D., Davoren, P., & Collett, A. (1997). Nuclear weapons test site cleanup: The Maralinga Rehabilitation Program. In: L. C. Mohr, W. Nixon, & D. Hoel (Eds), *International case studies in risk assessment and management* (pp. 59–80). Charleston, SC: The Medical University of South Carolina Press.

Deines, A. C., Goldman, D. I., Harris, R. R., & Kells, L. J. (Preparers). (1990). *Marshall Islands chronology 1944 to 1990*. Department of Energy. International Health Programs. Available at http://worf.eh.doe.gov:9090/ihp/chron/#alpha (accessed November 1997).

DEST (2004). The Maralinga Rehabilitation Project. Australian Department of Education Science and Technology. 12 March 2004. http://www.radioactivewaste.gov.au/managing_australia's_radioactive_waste/maralinga_rehabilitation_project/maralinga_faq.htm (Accessed May 2004)

Dorrance, J. C. (1992). The United States and the Pacific Islands. The Washington Papers No. 158. Center for Strategic and International Studies. Praeger Publishers. Westport, CT. 192pp.

Feinberg, R. (1995). *Seafaring in the Contemporary Pacific Islands: Studies in Continuity and Change*. DeKalb: Northern Illinois University Press.

Grabosky, P. N. (1989). Wayward Governance: Illegality and its Control in the Public Sector. Australian Institute of Criminology, Canberra http://www.aic.gov.au/publications/lcj/wayward/ch16.html (Accessed December 2001).

Guyer, R. L. (2001). Radioactivity and Rights: Clashes at Bikini Atoll. *American Journal of Public Health, 200; 91*(9, September), 1371–1376.

Jennings, A. B., & Leschine, T. M. (2000). The Bikini Atoll experience: Inherent fallibility of institutional controls and the virtues of "Defense in Depth". *Long Term Institutional Management of US Department of Energy Legacy Waste Sites* (pp. 54–55). Washington, DC: National Academy Press.

Kiste, R. C. (1968). Kili Island: A Study of the Relocation of the Ex-Bikini Marshallese. Project for the comparative study of cultural change and stability in displaced communities in the Pacific. No. 8. Published by the Department of Anthropology. University of Oregon. Eugene, Oregon. 393 p.

Kiste, R. C. (1974). *The Bikinians: A Study in Forced Migration*. Menlo Park, CA: Cummings Publishing Company.

Kristof, N. D. (1997). An atomic age of Eden (but don't eat the coconuts). New York Times (Late Edition (East Coast)) New York, NY. March 5 1997. p. A4.

LaPorte, T. R., & Metlay, D. S. (1996). Facing Deficit of Trust: Hazards and Institutional Trustworthiness. *Public Administration Review, 54*(4), 341–347.

LLNL. (2004). Internal Dose Monitoring: Enewetak Atoll. Lawrence Livermore National Laboratory. http://eed.llnl.gov/mi/enewetak.php (Accessed May 2004).

Maralinga Rehabilitation Technical Advisory Committee (MARTAC). (2003). Rehabilitation of Former Nuclear Test Sites at Maralinga and Emu (Australia) 2003. Department of Education, Science and Training, Commonwealth Government of Australia, Canberra.

McEwan, A. C., Simon, S. L., Trott, K.-R., Sankaranarayanan, K., & Paretzke, H. G. (1997). Some reflections on the role of the scientific advisory panel to the Marshall Islands Nationwide Radiological Study. *Health Physics., 73*, 265–269.

Minchin, N. (2000). Maralinga Rehabilitation Project: Response to certain issues raised by Radio National's Background Briefing. *Senator Nick Minchin, Minister for Industry Science and Resources, 1*(May).

Mohr, L. C., Nixon, W., & Hoel, D. (1997). *International Case Studies in Risk Assessment and Management*. Charleston, SC: The Medical University of South Carolina Press.

National Research Council. (2000). *Long-Term Institutional Management of U.S. Department of Energy Legacy Waste Sites*. Washington, DC: National Academy Press.

Niedenthal, J. (2001). *For the Good of Mankind: A History of the People of Bikini and their Islands*. 2nd edition. Majuro, Marshall Islands: Bravo Publishers. 264p.

Noshkin, V. E., & Robison, W. (1997). Assessment of a radioactive waste disposal site at Enewetak Atoll. *Health Physics., 73*, 234–246.

Nuclear Claims Tribunal. (2004). Marshall Islands Nuclear Claims Tribunal Website. Available at: http://www.tribunal-mh.org (accessed May 2004).

O'Conner, A. (1997). Special article coordinated by Alan O'Connor, Department of State Aboriginal Affairs, with assistance from the Aboriginal landholding authorities. Article reproduced from South Australian Year Book, 1997 (ABS Catalogue no. 1301.4).

Parkinson, A. (2002). Maralinga: The Cleanup of a Nuclear Test Site. *Medicine & Global Survival, 7*(2), 77–81.

Perrine, D. (2001). Atomic coconuts and remote control sharks: at the graveyard of the Pacific Fleet. *Skin Diver Magazine, 50*(4), 88.

Pockley, P. (2000). Clean-up strategy at Australian nuclear site called into question. *Nature, 404*. 20 April 2000. Available at http://www.nature.com (accessed December 2004). 797 p.

RMI Online. (1997). Nuclear Testing in the Marshall Islands: A chronology of events. http://www.clark.net/pub/rmiemb/nuccron.html (Accessed December 1998)

Robison, W., & Stone, E. L. (1992). The effects of potassium on the uptake of 137Cs in food crops grown on coral soils: Coconut at Bikini Atoll. *Health Physics*, *62*, 496–511.

Simon, S. L. (1997). A brief history of people and events related to atomic weapons testing in the Marshall Islands. *Health Physics.*, *73*, 5–20.

Smith, H. (1991). Rehabilitation of the former nuclear test sites in Australia. *Report of the Technical Assessment Group. J. Radiological Protection,*, *11*(4), 267–269.

Technical Assessment Group (TAG). (1990). Rehabilitation of Former Nuclear Test Sites in Australia, Report by the Technical Assessment Group. Australian Government Publishing Service, Canberra.

Thiess Services. (2003). Case Study: Maralinga Nuclear Test Site Rehabilitation. http://www.thiess-services.com.au

Weisgall, J. M. (1994). *Operation Crossroads: Atomic Tests at Bikini Atoll.* Maryland: Naval Institute Press.

THE CHALLENGES AHEAD

VULNERABILITIES AND UNCERTAINTIES IN LONG-TERM STEWARDSHIP

Curtis Travis

ABSTRACT

The need for long-term stewardship for hazards is widespread and increasing. Many of the largest and most complex sites requiring stewardship are within the U.S. Department of Energy's nuclear weapons complex, though many "brownfields" and Superfund sites in the private sector have similar needs for long-term care. High cleanup costs and difficult-to-resolve technical challenges are the main reasons why such sites are not being cleaned up to support unrestricted use. EPA policy changes are now redirecting cleanup efforts toward waste isolation and containment-in-place rather than waste removal. This shift is not being matched by a corresponding shift in the conceptual and operational approach to remediation planning. The process logic of cleanup continues to be "single-pass" and linear, via sequential planning and implementation stages that aim to declare sites "cleaned up and closed" once the sought-after degree of waste isolation from the biotic environment is achieved. Instead an iterative and nonlinear process logic is required – one that directs cleanup toward the needs of effective stewardship of the remaining residual contamination rather than toward regulatory standards that support the selected future land use for the site. This redirection of

Long-Term Management of Contaminated Sites
Research in Social Problems and Public Policy, Volume 13, 195–212

ISSN: 0196-1152/doi:10.1016/S0196-1152(06)13008-2

cleanup logic shifts the focus in remediation planning to vulnerabilities and uncertainties associated with long-term stewardship rather than the degree of isolation of wastes from human populations and the environment achieved. Such a shift is needed in recognition that situations of restricted human use of still-contaminated sites are being created that remain vulnerable to failure given the length of time over which stewardship is required.

INTRODUCTION

It is estimated that there are more than 200,000 highly contaminated hazardous waste sites in the United States (about 170,000 of them in the private sector), and that cleaning them up will require an estimated $300–$700 billion (National Research Council (NRC), 1994, 2003a). The worst of these sites, in terms of complexity and cost, are not the industrial sites, as was first believed, but sites owned by the federal government. For example, the US Department of Energy (DOE), as a result of nuclear weapons production and research programs, has contaminated over 140 sites, some quite large, with high levels of chemical and radioactive waste. The Department of Defense (DOD) has more than 23,300 sites contaminated with unexploded ordnance, heavy metals, jet fuels, solvents, and chemical weapons waste.

It is increasingly recognized (DOE, 1999; GAO, 2002, 2003; NRC, 2003a) that it is neither technically nor economically feasible to restore most of these contaminated sites to a pristine condition, that is, to a point where they can be released for unrestricted use (NRC, 2000, 2003a). Even after significant cleanup, residual hazards will remain, both on the surface and in the subsurface. In many cases, radioactive wastes that remain at sites will pose hazards far longer than the life of the engineered and institutional controls designed to contain them and these hazards will need to be managed for hundreds or thousands of years. Thus, the completion of active remediation does not mean the end of environmental responsibilities. In recognition of this fact, DOE and DOD acknowledge the need to incorporate a long-term stewardship perspective into their remediation decisions (DOE, 1999) although little action has been taken toward realizing this goal.

Stewardship consists of those engineering, institutional, and administrative activities that will be required to protect human health and the environment from hazards that remain at a site after the cessation of remediation activities (NRC, 2000, 2003b). Stewardship activities will be required as long as the hazards exist, which in the case of radioactive waste may be nearly infinitely. DOE acknowledges that long-term stewardship will be needed at more than 100 sites encompassing more than 350,000 acres of land throughout the United States (DOE, 2001; NRC, 2000).

Despite general agreement that it is often necessary to leave residues at waste sites, there are significant scientific, technical, and societal uncertainties associated with doing so. Most sites where long-term stewardship will take place will not be completely characterized. Often waste is "capped in place" without knowledge of the true nature or extent of the subsurface contamination. In addition, subsurface hydrological conditions are generally not well understood. Natural and engineered barriers are expected to hold safe these hazardous materials for many hundreds of years after the site is "cleaned up and closed." However, without knowing the extent or potential mobility of the subsurface contamination, nor the reliability of the containment technologies, uncertainties abound.

Stewardship efforts thus face several challenges: vulnerabilities associated with barrier failure, institutional change over multigeneration time periods, and uncertainties associated with the inability to precisely predict future factors that will affect exposure and risks. In this paper, we discuss a framework for identifying and addressing some of the vulnerabilities and uncertainties associated with stewardship.

THE CURRENT CONCEPT OF STEWARDSHIP IS LINEAR

The current concept of stewardship as practiced by most federal agencies (DOE in particular) is linear. In the linear model, cleanup standards are selected based on some proposed future land use, the site is cleaned to these standards to the extent possible, and stewardship needs are defined by the level and nature of contamination remaining after "cleanup" is achieved. This model is called linear because it proceeds in a linear fashion from establishing future land use through site characterization, remedy selection, cleanup, and finally to defining stewardship needs and stewardship implementation (see Fig. 1). "Linear staging is a single predetermined path defined primarily by milestones driven by program schedule, cost, and technology" (NRC, 2003b).

DOE guidance (DOE, 2001) for stewardship suggests the following steps: (1) Identify possible future uses of land and water at the site, (2) Select an end-state consistent with these uses, (3) Select cleanup standards consistent

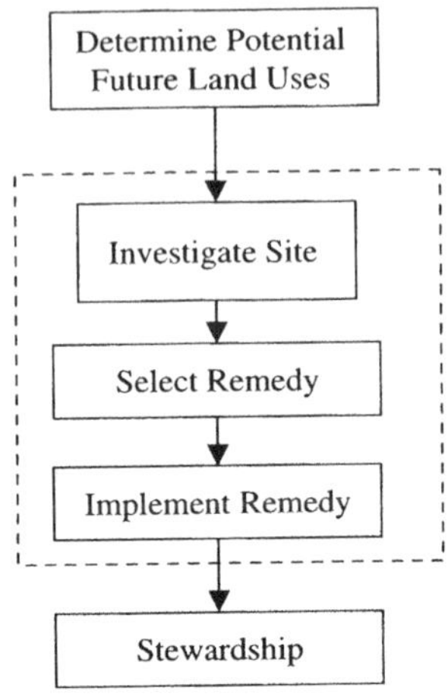

Fig. 1. Current Remediation Process.

with the end use, (4) Remediate the site, and (5) Place remaining contamination in stewardship. Thus the DOE view of stewardship is that stewardship begins where remediation ends (DOE, 2001). Once a site is remediated to the extent possible, taking into account cost and technology, the remaining waste is placed into stewardship. The level of cleanup achieved defines the needs of stewardship.

A weakness of this approach is that the vulnerabilities and uncertainties associated with long-term stewardship needs are not directly and explicitly accounted for during planning for remedial actions. The current EPA and state regulatory focus is on rapid and economical contaminant reduction, and significant gains have been made in that direction. Hot spots in surface soils are being excavated, as are pits, trenches, and settling ponds. Contaminated buildings are being cleaned or demolished. But typically large amounts of subsurface contamination will remain. Measures are being taken to contain the subsurface contamination at some sites, but the long-term likelihood of success of these measures is unknown.

While DOE frequently states that future land uses determine the end-state conditions that must be achieved during cleanup, in reality the process is often the reverse. It is frequently the case that economic, technical, or worker safety considerations limit the end-state conditions that can be achieved, and thus limit future possible land uses. It is for this reason that the most frequently proposed future land use at DOE facilities is as an industrial park (Hanford, Washington), a wildlife reserve (Rocky Flats,

Colorado), or an "ecologically restored" area (Fernald, Ohio). In these cases, subsurface soils and groundwater do not have to be fully restored, and large surface soil areas may be cleaned to only a limited extent. Thus, in reality, cost considerations are placing limitations on the amount of cleanup achievable and thus on potential end uses and not the other way around, as is frequently stated.

THE NEED FOR STEWARDSHIP IS INCREASING

The US Superfund program embodies well the change in philosophy that is leading to less site cleanup and greater reliance on long-term stewardship. The original promise of Superfund was to clean up toxic waste sites to a point where they no longer posed a hazard to human health or the environment, and could thus be released for unrestricted reuse. The Superfund legislation emphasized the need to use permanent solutions and urged a bias toward treatment and recovery to the maximum extent practicable (EPA, 1997). (Treatment is any process that changes the physical, chemical, or biological character of a waste to make it less of an environmental threat.)

It is now recognized, however, that this goal is not obtainable for most hazardous waste sites, particularity those complex sites found at DOE and DOD facilities. Beginning in 1993, EPA began to implement a series of reforms that in part address this issue. EPA (1997) issued guidance as to when exhumation and removal of wastes versus stewardship should be used. To facilitate the distinction, EPA separated source terms at a site into two categories: principal threat and low-level threat sources. "Principal threat source terms are those source materials considered to be highly toxic or highly mobile which generally cannot be contained in a reliable manner or would present a significant risk to human health or the environment should exposure occur" (EPA, 1997). "Low-level threats are those source materials that generally can be reliably contained and that would present only a low risk in the event of exposure" (EPA, 1997). EPA suggests that treatment, broadly construed, be used to address the principal threats posed by a site wherever practicable, and that containment be used to address low-level threats or situations where treatment is impracticable EPA (1997).

The major problem at hazardous waste sites that defies current technological solution is contaminated subsurface soils and groundwater that are more than 6 ft underground. In an effort to clean these media to safe levels, "pump-and-treat" and other in situ contaminant extraction systems (e.g., vapor extraction) have been applied. Pump-and-treat systems have demonstrated some success at removing contaminant mass from a groundwater plume, reducing hot spots within a plume, or achieving containment of a plume (NRC, 1994). However, they have not been shown effective at restoring aquifers to drinking water standards (Travis & Doty, 1990). Removing contamination from subsurface soils and groundwater is a costly and time-consuming process, which to date has demonstrated little progress. Much of the regulatory guidance for groundwater cleanup was written before 1989 when the limitations of pump-and-treat systems were not fully appreciated (DOD, 1998; EPA, 2003). Since the Superfund Law provides no criteria for termination of these systems (other than total cleanup of the soils and groundwater), they can stay in place indefinitely.

At a national level, the cost of installation and maintenance of pump-and-treat systems is reaching into the billions of dollars with little demonstrated return. In many cases, these groundwater systems are not needed to confine the groundwater plumes onsite or to protect human health. The DOD currently spends $40 million annually maintaining existing pump-and-treat systems and estimates that the cumulative cost through 2020 (including planned additional systems) will be $2.3 billion (DOD, 1998). In response to the rising cost of cleanups and the acknowledged ineffectiveness of cleanup technologies, DOE and DOD have been advocating containment rather than cleanup.

The Navy has about 4,600 contaminated sites. After spending $3.8 billion to remediate these sites, 60% have reached the "response complete" stage (NRC, 2003a). However, target cleanup levels at these sites have not been reached. That is, the Navy has done everything the EPA and the states have asked then to do, and most sites are still not clean. The Navy estimates that it will take another $4.8 billion to reach the "response complete" stage on all of its sites. This problem is not unique to the Navy. At the end of fiscal year 2002, 47% of Superfund sites were in the "construction complete" stage (GAO, 2003). Construction complete means that "physical construction of all cleanup actions is complete and that all immediate threats have been addressed, and all long-term threats are under control" (EPA, 2001). Most of these sites are not cleaned up to unrestricted use standards, however.

Partly in response to the high cost of removing contaminants from subsurface soils and groundwater (and largely without explicit acknowledgement), the EPA has been gradually changing the goal of Superfund. Instead of removing contamination to levels compatible with unrestricted human use, the new goals are to control human exposures and to prevent remaining contamination at the site from moving off site. The percentage of Superfund sites selecting pump-and-treat as a remedy for contaminated groundwater peaked in 1993 at 74% and declined to 47% in 1999 (EPA, 2001), as the goals of Superfund have changed from removal to containment. In contrast, the percentage of sites using institutional controls, monitoring and other remedies has steadily increased with the realization that pump-and-treat does not work. Institutional controls are now a part of the remedy at 49% of Superfund sites (EPA, 2001). Remedies relying on monitored natural attenuation (i.e., allowing contaminants to abate via such natural processes as evaporation, dispersion, or, in the case of radioactive substances, radio-active decay) have increased from 0% in 1983 to 25% in 1998.

The Love Canal Superfund site, a former chemical dump that was the igniting force in passage of the Superfund legislation, is an example of this new trend. In March of 2004, the EPA declared Love Canal "cleaned up and closed" and removed Love Canal from the Superfund list. However, 21,000 tons of chemical waste remain on-site. The cleanup at Love Canal has centered on containing the waste under a 40-acre clay cap and surrounding it with a barrier drainage system. Two hundred nearby formerly boarded-up homes have been renovated and sold to new owners (at a reduced price), and 10 new apartment buildings have been constructed near the site. This is not to say that the contained waste at Love Canal represents a risk to surrounding populations. It does point out that it may be impossible to use permanent treatment and recovery as the primary approach to remediation of hazardous waste sites, and that "cleaned up and closed" sites may have tons of hazardous waste remaining on site in need of continued stewardship.

The conundrum faced by the Superfund program is that on the one hand it can spend billions and still not be able to reach safe levels or, on the other, it can leave waste in place and possibly expose future generations, even though able to establish safe conditions today. The evolution of the Superfund program from a focus on permanent solutions using treatment and recovery to a program largely focused on containment and long-term isolation of radioactive and hazardous waste raises questions as to the adequacy of stewardship programs; these questions are not being adequately addressed. As EPA (1997) correctly points out, containment can safely be used for "low-level threats that generally can be reliably contained and that would present only a low risk in the event of exposure." Unfortunately, most hazardous waste sites and, in particular, most DOE sites do not fall in this category. Containment is now being used as the method of choice for the most complex remediation problems, even sites that are "highly toxic or highly mobile and would present a significant risk to human health or the environment should exposure occur" (EPA, 1997). This situation highlights the need for closer attention to stewardship vulnerabilities and uncertainties. The goal of stewardship is hazard containment, while the focus of Superfund has long been on the ability of technologies to remove wastes. Technologies applied in stewardship must be able to isolate wastes, and the two goals are fundamentally different.

SUPERFUND REMEDIATION AND STEWARDSHIP HAVE DIFFERENT GOALS

Nothing illustrates the differences between remediation and stewardship better than the current evolution of the use of the term "cleaned up and closed." DOE often states that it will produce "cleanup and closure" of its sites by a given date (e.g., Rocky Flats by 2006). But cleanup and closure does not mean returning the site to a condition of unlimited use. Difficult-to-estimate quantities of radioactive and hazardous chemical wastes will remain in subsurface soil. The US Congress, aware of the high cost of remediating federal sites nationally, is supporting this evolution in meaning. As defined by the US Congress, the term "cleanup and closure" means "the response actions for covered substances have been carried out as required by Federal and state agencies" (U.S. Congress, 2002). This twisting of words indirectly weakens the stewardship program. Calling the sites closed gives the impression that there is nothing left to do. The use of the terminology "cleaned up and closed" minimizes perception of the hazard of the remaining waste and causes the average citizen to believe that all remaining hazards have been removed.

Consider the case of Melton Valley, a former waste disposal area of Oak Ridge National Laboratory, scheduled for "cleanup and closure" by 2006. The Melton Valley was used between 1951 and 1986 for disposal of approximately 2 million curies (Ci) of radioactive and mixed waste, including uranium and transuranic wastes (DOE, 2002a), and is the source of 90% of the strontium and tritium discharged from DOE property into the Clinch River, a source of drinking water and a recreational water body. In places there is only 0.6 m of separation between the buried waste and the groundwater table, and travel times for contaminates to reach the Clinch River following a heavy rain are on the order of days. The closure plan calls for the wastes to be "contained in place" under 125 acres of caps. Collection drains will be needed downgradient of capped areas to collect contaminated leachate since all water flow through the area cannot be stopped. Contaminated surface soils with activity levels of less than 2,550 μR/h will be left in place. Moreover, the waterway system that drains the Melton Valley – the White Oak Creek and White Oak Lake system – is not included in the closure plan and will still be awaiting mitigation after 2006. An estimated 700 Ci of ^{137}Cs and 325 Ci of ^{60}Co were released from White Oak Lake into the Clinch River between 1949 and 1992 (DOE, 1988). In a 1983 letter from the state of Tennessee to Oak Ridge officials, White Oak Lake is described as "the most radioactive lake in the country." Thus, the Melton Valley will hardly be "cleaned up and closed" by 2006; so why use this terminology? In fact, surface water will not meet drinking water standards until ten years after all closure elements are in place, through natural decay of the residual contaminants (DOE, 2002a). The misleading nomenclature of "cleaned up and closed" obscures the need for effective long-term stewardship.

At Hanford contaminated subsurface soil and groundwater will remain a problem after "cleanup and closure." Over 450 billion gallons of liquid contaminated with radionuclides and chemicals have been discharged or leaked into soils over the years (DOE, 2002b), including more than a million gallons of high-level radioactive waste from 177 underground tanks. It is anticipated that groundwater contamination under the site will exceed the drinking water standards for 150–300 years following "cleanup and closure." The 100N area includes two of the most contaminated "cribs" (holding ponds and trenches into which liquid radioactive wastes were discharged) on the Hanford Site. Located along the Columbia River, the N area remediation effort involves removing 150,000 t of soil and debris from these cribs that formerly held contaminated water from the N Reactor's cooling system and fuel basin. They operated in much the same way as a septic tank and drain field and are now up to 50 times more radioactive than other sites at Hanford. After removal and disposal of the top 15 ft of soil from the waste site, DOE is proposing to leave the rest of the contaminated soil in place with institutional controls to ensure public protection. However, beneath these soils, radioactive strontium-90 has affected about a square mile of Hanford groundwater that is currently entering the Columbia River at an area known as N Springs. The existing pump-and-treat system at 100N has proven to be an ineffective option for long-term remediation (DOE, 2002b). What is needed, but what has yet to be spelled out in sufficient detail, is the system of controls (i.e., long-term stewardship) that will protect the public and the resources of the Columbia River from harm from the remaining wastes over the time period necessary, and how the effectiveness of that system will be maintained over the period of its need.

The uncertainties and vulnerabilities of stewardship are important issues that must be explicitly acknowledged and addressed in a proactive fashion, and not marginalized by implying that sites are "cleaned up and closed" (in the literal sense of the phrase), and, therefore, presumed to present no further challenges.

VULNERABILITIES ARE THE DRIVERS OF STEWARDSHIP NEEDS

Vulnerability in a long-term stewardship context can be thought of as the susceptibility of humans or the environment to the residual risks associated with contaminants left in place at sites under long-term stewardship. While vulnerabilities are most clearly associated with potential failures in engineered, institutional, or administrative measures to contain contamination, significant uncertainties exist regarding just what could result in loss of protection. The current view is that residual contamination remaining at a site is the post-cleanup driver for how extensive stewardship measures must be. In reality, the extensiveness of stewardship measures must be determined before remediation commences by evaluating before-the-fact vulnerabilities inherent in the end state of the site. For simple sites with limited contamination, the distinction between vulnerability and risk may not be significant. But for complex sites with extensive subsurface contamination, the distinction is critical. The primary distinction between the two cases lies in the ability to correct future failures that may occur.

The primary vulnerabilities that must be evaluated in stewardship systems are:

- *Barrier effectiveness.* Demonstrating how a combination of geologic (natural) and engineered (man-made) barriers will operate effectively over long periods of time.
- *Predicting performance.* Understanding and predicting how institutional controls and other institutional management measures will perform over long time periods.
- *Monitoring performance.* Evaluating the performance of the stewardship system itself relative to pre-established standards.

The magnitude and nature of vulnerability should determine the extensiveness of the remediation activities required. High-consequence events require robust designs and redundancies to minimize chances of their occurrence (and situations with very high-consequence events as potential outcomes should not be left as residual risks). Events with short lead times also require robust designs and redundancies. Event with long lead times are best addressed as they develop, but first have to be recognized. Computer models can assist with anticipating such events, but it is difficult to predict the performance of a stewardship system over long periods of time. To determine if safety is compromised by certain vulnerabilities, sensitivity analysis can be conducted in which the consequences of deviations are evaluated. "Observations of the natural systems at the site can play an important role in evaluation of the systems because such systems have evolved over long-time scales" (NRC, 2003c). Prioritization of possible deviations depends on such characteristics as the likelihood of occurrence, severity of the impact resulting from the deviation, the ability to detect a barrier failure, the time that will be available to respond to the deviation, and the ability to correct the deviation.

The major vulnerabilities in stewardship define its science and technology research needs. The top priorities are barrier performance, subsurface science, early detection of barrier failure and enhanced ability to respond to it, and methods to improve the long-term performance of institutional controls. The focus of stewardship on hazard containment reduces the consequence of limited knowledge of toxicology and uncertainties in risk assessment. The major uncertainty left is predicting the long-term performance of barrier systems, whether engineered or institutional and administrative.

THE STEWARDSHIP CONCEPT SHOULD BE NONLINEAR AND ITERATIVE

Following stabilization of contaminated areas, barriers will separate the remaining hazards from key receptor groups (e.g., workers, the public, and the environment). These barriers may be engineered features (to stabilize and/or contain or isolate waste) or institutional/administrative controls (to restrict certain uses and hence exposures, provide important stewardship information, or maintain appropriate security). In many cases, however, radioactive and hazardous chemical wastes that remain will pose hazards for far longer than the life of the original barriers designed to contain them. The finite life span of engineered solutions and institutional controls presents special challenges for long-term protective maintenance. It is unlikely that any nation can design an initial system that will surmount these challenges and successfully isolate hazardous wastes for the thousands of years that may be required. "Selection of barriers must allow future adaptations to be carried out without unacceptable impact on safety" (NRC, 2003c).

Three approaches exist for handling the long-term uncertainties associated with stewardship. One is the stewardship approach currently taken in the cleanup of most complex sites. This approach is to pay minimal attention to vulnerabilities and hope that nothing unforeseen happens in the future. The unknown consequences of present actions are left for a future generation of solve. This approach to stewardship can be called "passively adaptive." It is adaptive because trajectory modifications can be made in the future. But it is passive because no precautions are taken during planning for stewardship to ensure that future problems, should they occur, can be corrected. It is based on the presumption that stewardship practices implemented today will not themselves lead to irreversible harm, so that adaptive changes can be made in the future if problems arise. However, no analysis is done when measures are being selected to assure the validity of this assumption.

An alternative approach (at the other end of the spectrum) is to completely understand the science of a site and build barriers that cannot possibly fail, given that understanding. This approach is based on extensive site characterization, combined with computer modeling that predicts conditions far into the future. Barriers and other safeguard systems are then designed on the basis of these predictions. (This is the approach being taken at the high-level radioactive waste repository at Yucca Mountain.) The disadvantage of this approach is that it is virtually impossible to perform analyses and measurements today that will reduce to acceptable levels uncertainties over hundreds of years (witness the controversy over siting the repository at Yucca Mountain). Since uncertainties cannot be eliminated when the system is initiated, barrier systems will have to be overdesigned, greatly increasing the costs. Moreover, no current technological solution will be adequate without modification for hundreds to thousands of years.

A third approach and the approach advocated here, is to consider future vulnerabilities when selecting barriers, place reasonable bounds on uncertainties through an iterative process that seeks to maintain barrier integrity and institutional controls for a generation (20–50 years) and then passes responsibility to the next generation. This approach to stewardship is called "actively adaptive." The National Academy of Sciences defines adaptive management as "a continuous and deliberate decision-making and management process … that emphasizes continuous learning, both technical and societal, includes scientific and managerial re-evaluation and reactions to new knowledge, is responsive to stakeholder input, and is designed to

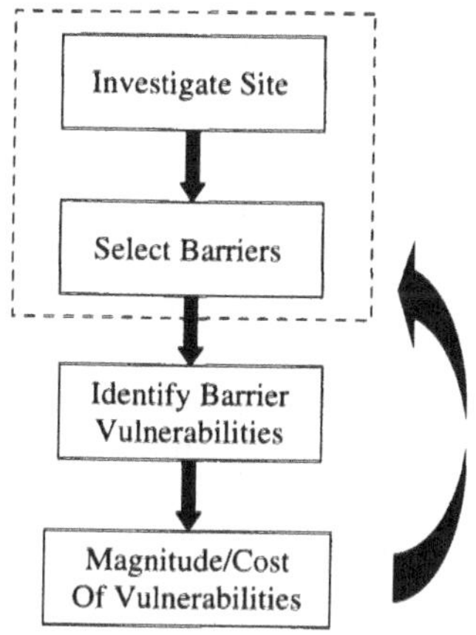

Fig. 2. Actively Adaptive Stewardship Process.

continually improve the project while retaining the option of reversibility as much as possible" (NRC, 2003c).

Under this approaçh, the major factor driving selection of remediation options at a site is not isolation of the waste from current populations, but rather ensuring that barrier failures will not lead to irreversible damage in the future. This is a question that must be addressed before remediation plans are finalized. Vulnerabilities are the drivers of cleanup actions, not land end uses, as DOE currently proposes. Guidance for stewardship should consist of the following steps (see Fig. 2): (1) identify critical receptors to be protected by stewardship, (2)select possible barriers, (3) identify barrier vulnerabilities and the possibility for irreversible harm, (4) determine cost of reducing vulnerabilities, (5) select new barriers or more extensive remediation, and (6) iterate the process. Thus, contrary to the DOE view that stewardship begins where remediation ends (DOE, 2001), remediation cannot begin until stewardship vulnerabilities are identified and controlled.

In an evaluation of the DOE cleanup program, Russell (2002) points out that (1) large quantities of long-lived radioactive wastes at DOE site are a reality, and cannot be made to go away, and (2) it is technically impossible to eliminate the risks generated by these wastes or to guarantee that the contamination can be contained for long periods of time. Russell noted that there is a tendency for DOE "to separate artificially the initial stage of waste management – cleanup of the sites – from that which will come later – stewardship." Russell concluded that the focus of DOE (and the Congress) on minimizing current cost is not in the best long-term interest of the United

States, and that DOE should redirect its environmental program. "An important element in that redirection is that DOE should integrate future consequences into its current decisions – there can be no separation between present and future costs and risks. "The decision process should start with an unbiased evaluation of expected actual consequences" of current remediation decisions (Russell, 2002).

The major difference between "actively adaptive" and "passively adaptive" stewardship is the attention paid during selection of remediation remedies to the question of irreversible consequences arising from current actions. Both approaches advocate leaving subsurface contamination in place with barriers. But actively adaptive management accounts for vulnerabilities and uncertainties during the design of stewardship measures. It does not make the presumption that future failures in barrier technologies will be reversible. Instead, barrier technologies are designed to insure that failures will have reversible consequences. This approach assumes that stewardship will require continuous management and dynamic responses to changing future conditions. The principle of this approach is to address uncertainty through further site investigation and possible development of enhanced barriers. However, it does not strive to eliminate all major uncertainties prior to beginning stewardship; instead, it manages potential deviations during stewardship through aggressive monitoring and contingency planning.

Ultimately, the challenge of stewardship is to implement a continuous framework that ensures that hazards are contained, appropriate monitoring and contingency plans are in place, and knowledge regarding existing hazards is communicated to future generations. A recommended hazard management framework for stewardship is as follows:

- *Divide stewardship into discreet, manageable time units (20–50 years)*. It is impossible to plan for stewardship needs over the near infinite time periods that may be needed. It is widely recognized that long compliance periods introduce irresolvable uncertainties and that stewardship requires time periods that may be longer than institutional controls or monitoring programs might be effective. Thus, it is necessary to focus on the short term and do what is necessary to contain the contamination, increase understanding of the site, and pass it on to the next time period in the same or better condition than received. The multigeneration uncertainty associated with long-term hazards can be managed through an iterative process of managing site conditions over successive 20–50 year intervals. Barriers and monitoring systems should be designed to contain contaminants over these time periods.

- *Define expected conditions.* On the basis of current knowledge, identify the most probable sequence of events over the next 100 years (two generations), assuming gradual barrier degradation. This analysis can be both qualitative (expert opinion) and quantitative (computer models). To assist in development of contingency plans, perform general trending analysis for a 1,000-year period.
- *Identify plausible deviations from a controlled trajectory.* Evaluate vulnerabilities of the barriers separating public, worker, and environmental receptors from site hazards to provide a qualitative evaluation of the likelihood and consequences of barrier failure. Examples of potential vulnerabilities include degradation of isolation systems, failure of administrative controls, lack of understanding of subsurface environment, degradation of packaging, criticality, and the presence of contamination outside containment systems. Next, accounting for uncertainty in current knowledge, develop reasonable bounding estimates of the consequences of plausible deviations.
- *Eliminate deviations that would result in loss of control.* Perform additional site investigations and actions sufficient to eliminate or at least reduce deviations that lead to unacceptable risks. This may require installation of additional barriers or removal of additional waste. High-consequence events require robust designs and redundancies, as do events with short lead times. The iterative investigation process can be stopped when remaining uncertainties are unlikely to lead to unacceptable loss of control.
- *Develop contingency plans for plausible deviations.* Develop contingency plans for detecting and responding to plausible deviations from a controlled status. Evaluate the cost, design, construction, detection, and contingency challenges presented by these deviations. The cost of controlling future deviations needs to be considered when selecting the current remediation strategy. It is the combined short- and long-term cost of remediation, and securing the site for future generations must be considered in selection of current remediation approaches. Minimizing current costs at the expense of increased future vulnerabilities is not acceptable. The need for continued vigilance in monitoring for future deviations from the expected trajectory requires a parallel development of support (e.g., funding) mechanisms.
- *Monitor for deviations.* Institute a rigorous surveillance and maintenance program to ensure that barriers remain reliable and surrounding site conditions remain safe. Specific performance and monitoring objectives should be developed on the basis of probable conditions and reasonable

deviations. Action thresholds should be identified that define the maximum deviation that will be tolerated before contingency plans are implemented. Again, the need for continued monitoring requires development of support mechanisms. It is unreasonable to assume that monitoring will be performed continuously far into the future. Thus, a stewardship program cannot rely on monitoring alone to assume that the program is operating as planned. Monitoring is worthwhile and needed in the short term to increase knowledge of how the barriers are functioning and to assure that contamination is not moving. Predicting long-term stewardship performance requires reliance on mathematical modeling.

- *Improve understanding of site over time and manage iteratively.* Stewardship is an active process that requires continual monitoring, analysis, and improved understanding over time. Confidence in long-term safety of the system should be maintained and enhanced over time. Stewardship must relay on continuous learning throughout the program. Increasingly advanced levels of scientific and technological knowledge should be incorporated by each generation to produce increasingly safe containment, so that each generation can pass the site to the following generation in at least as good a condition as that when it was received. Accomplishing this would also require access to funds or other forms of support as needed.

CONCLUSIONS

Since its implementation in 1980, the goals of the Superfund program have been slowly evolving, as it has had to face some hard realities. The original goal of Superfund was to protect human health and the environment by restoring hazardous waste sites to a pristine condition, that is, a condition consistent with unlimited use. However, the inability of current technical solutions to reach these levels and the rising cost of remediation are motivating many to advocate containment rather than cleanup. Rather than focus on returning sites to an unrestricted end-use state (which in most cases may be impossible), they advocate focusing on source control and limiting current and future risks. However, this strategy leaves significant contamination in the environment.

Any containment strategy must be accompanied by long-term stewardship to insure that the public is protected over the long term. The current approach to stewardship is not adequate. It relies on removing or controlling current sources of contamination and hoping that these measures will be adequate over the long term. No analysis is performed to identify the consequences of failure. It is assumed that if problems develop in the future, there will be time to address them and that the damage will be reversible. But this is only an assumption because no analysis has been done to verify this belief.

An alternate approach is to focus on vulnerabilities, rather than the level of remaining contamination. Vulnerabilities result from failures in engineered, institutional, or administrative barriers to contain contamination. The level of vulnerability at the site determines the level of stewardship needed. High-consequence events require robust designs and redundancies, as do events with short lead times.

Active stewardship assumes that failures in engineered barriers may lead to irreversible damage. Thus, before remedial alternatives are selected, analysis is performed to identify the most probable sequence of events over the next 100 years (two generations), assuming gradual barrier degradation. Next, accounting for uncertainty in current knowledge, reasonable bounding estimates are developed of the consequences of plausible future deviations. Consequences that pose insurmountable future challenges from a technical or cost perspective must be eliminated through the current installation of additional barriers or removal of additional waste.

Stewardship must be an active process that utilizes continual monitoring and analysis to improve understanding over time. Confidence in long-term safety of the system should be maintained and enhanced over time. Stewardship must relay on continuous learning throughout the program (NRC, 2003c). The aim is to pass the site on to future generations in the same or better condition than received.

In summary, the traditional concept of hazardous waste site cleanup in the United States is that of a linear process, beginning with the determination of possible future land uses and proceeding along the optimal path constrained by cost and technology to eventual site cleanup. Stewardship needs are defined by the nature and extent of residual contamination left at the site following cleanup. The vulnerabilities and uncertainties associated with long-term stewardship are not directly and explicitly accounted for during current planning for remedial actions.

Instead, stewardship planning should be an "actively adaptive" process. It is active because precautions are taken during remediation planning to ensure that problems, if they occur within a generation, can be cost-effectively corrected. It is adaptive since it does not attempt to reduce uncertainty to zero, but rather designs stewardship controls (barriers) so trajectory modifications can be made in the future. It is designed to place reasonable bounds on uncertainties through an iterative process that seeks to maintain barrier integrity and institutional controls for a generation (20 to 50 years) and then passes responsibility to the next generation. Thus, barrier vulnerabilities should be the driver of cleanup actions, not site end-uses, as is the current practice.

REFERENCES

Department of Defense (DOD). (1998). *Evaluation of DOD waste site groundwater pump-and-treat operations*. Office of the Inspector General (98-090).

Department of Energy (DOE). (1988). *Historical releases from current DOE Oak Ridge operations office facilities*.

Department of Energy. (1999). *From cleanup to stewardship*. DOE/EM-0466.

Department of Energy. (2001). *Long term stewardship study*.

Department of Energy. (2002a). *Oak Ridge comprehensive closure plan*.

Department of Energy. (2002b). *Performance management plan for the accelerated cleanup of the Hanford site*. DOE/RL-2002-47.

Environmental Protection Agency (EPA). (2003). Office of Inspector General. *Improving nationwide effectiveness of pump-and-treat remedies requires sustained and focused attention to realize benefits*. 2003-P-000006.

Environmental Protection Agency. (2001). *Focus on construction completion. 757th completion*. EPA 540-F-00-023.

Environmental Protection Agency. (1997). *Rules of thumb for Superfund remedy selection*. EPA 540-R-97-013.

Government Accounting Office (GAO). (2002). *Waste cleanup: Status and implications of DOE's compliance agreements*. GAO-02-567.

Government Accounting Office (GAO). (2003). *Superfund program: Current status and future fiscal challenges*. GAO-03-850.

National Research Council (NRC). (1994). *Alternatives for groundwater cleanup*.

National Research Council. (2000). *Long-term institutional management of U.S. Department of Energy legacy waste sites*.

National Research Council. (2003a). *Environmental cleanup at navy facilities: Adaptive site management*.

National Research Council. (2003b). *Long-term stewardship of DOE legacy waste sites: A status report*.

National Research Council. (2003c). *One step at a time: The staged development of geologic repositories for high-level radioactive waste*.

Russell, M. (2002). *DOE legacy waste cleanup and stewardship: Beyond the top-to-bottom review*. JIEE 2002-06, Joint Institute for Energy and Environment, University of Tennessee.

Travis, C. C., & Doty, C. B. (1990). Can contaminated aquifers at Superfund sites be remediated?. *Environmental Science and Technology*, *24*(10), 1464–1466.

U.S. Congress. (2002). *Bill S.1438 establishing Rocky Flats as a national wildlife refuge*.

A 19-YEAR PERSPECTIVE ON LONG-TERM CARE ISSUES

Denise Bierley

ABSTRACT

This paper offers a personal perspective on the author's experience working with issues relating to the long-term management of nuclear contaminated sites, from the programmatic to the site-specific. Long-term care is and will be far more challenging than remediation activities; thus, the dynamics of long-term care require different approaches to problem solving. The need for nonlinear thinking will challenge management that has traditionally relied on linear approaches. Integrated risk management potentially offers some powerful and flexible tools for identifying and managing uncertainties. Managing uncertainties involves not only traditional budget, schedule, cost, and worker safety issues, but also other influences that are not easily quantifiable, including regulatory, cultural, social, political, legal, and "quality" issues. Understanding and incorporating changes in social context is critical to the planning and implementation processes of long-term care; the Department of Energy (DOE) must utilize processes that have consistency over time and that involve the public throughout the process. Management in the long term must reflect an understanding of how human systems function and how they couple with technological systems. DOE's relative success with its Uranium Mill Tailings Remedial Action Program exhibits some of these components.

Long-Term Management of Contaminated Sites
Research in Social Problems and Public Policy, Volume 13, 213–226

ISSN: 0196-1152/doi:10.1016/S0196-1152(06)13009-4

Many are now recognizing these components as key needs for any long-term care program for long-lived hazards.

This paper is intended to give a personal perspective related to long-term management of contaminated sites. From the mid-1980s until the mid-1990s, I worked as a contractor on the Department of Energy's Uranium Mill Tailings Remedial Action (UMTRA) Project, which has remediated 22 former uranium mill sites in 10 states. In 1996, I joined the National Academy of Sciences Committee on the Remediation of Buried and Tank Wastes, which was tasked with examining long-term institutional management issues related to Department of Energy (DOE) sites (NRC, 2000a). In my current position as a consultant to Sandia National Laboratories (Sandia) I work exclusively with DOE and Sandia on long-term care issues, specifically with helping to implement Sandia's long-term care program.

It is my intent to discuss a variety of topics and issues in this paper based on my observations and experiences with long-term care over the past 19 years. These range from the programmatic to the site-specific with the focus on the Department of Energy's programs.

LONG-TERM STEWARDSHIP IN CONCEPT AND PRACTICE

Long-term care, long-term management, surveillance and maintenance, stewardship, and institutional management have all been used to describe the diverse activities associated with management of various types of waste sites.

The DOE has defined "long-term stewardship" as "the physical controls, institutions, information and other mechanisms needed to ensure protection of people and the environment at sites where DOE has completed or plans to complete 'cleanup' (e.g., landfill closures, remedial actions, removal actions, and facility stabilizations). This concept of long-term stewardship includes, inter alia, land-use controls, monitoring, maintenance, and information management" (DOE, 2001a).

While this is DOE's definition of "stewardship," there are those that disagree with the concept of leaving waste in place and calling it stewardship. I like to call these disagreements semantic wars. The reality is that, for a variety of reasons, not all of DOE's contamination can be cleaned up to a residential risk scenario. Moreover, this is not just DOE's problem. It is a national problem that spans multiple agencies (e.g., Environmental Protection Agency and Department of Defense) and sites such as municipal solid waste facilities, Resource Conservation and Recovery Act (RCRA) hazardous waste disposal facilities, and Superfund sites. These facilities will also require varying degrees and types of institutional management activities to ensure protection of public health and the environment into the foreseeable future.

The institutional management systems that are required for long-term care of DOE sites will be the same ones needed to ensure the appropriate care of these other facilities across the nation. An example is the concept of robust institutional controls. Consideration of factors involved in ensuring appropriate institutional controls for application to DOE sites are the same ones needed to ensure that a former municipal solid waste facility is not inappropriately utilized, given its remaining contamination.

The DOE's emphasis on long-term care has been criticized as an attempt to avoid expensive cleanups. In addition, DOE has received criticism regarding the current administration's lack of funding for stewardship. While the UMTRA and other sites are managed within the DOE's new Office of Legacy Management, sites such as Sandia have had little program direction or funding from DOE headquarters. Beyond the obvious consequences, this makes for a fragmented approach to dealing with stewardship issues as well-intentioned people within the system try to piece together resources to address planning for long-term care. This is the current situation at Sandia.

Decision-Making Considerations

Much of what long-term management of contaminated sites entails is the management of long-term uncertainties. This can involve limited "programmatic" risk management approaches and the implementation of decision-making frameworks. I personally believe that the dynamics of long-term care require different approaches to problem solving. Integrated risk management potentially offers some powerful and flexible tools for managing uncertainties.

Risk management has many definitions. In the context of managing uncertainties, a very broad, perhaps nontraditional definition of risk management is required. Within the DOE and elsewhere, I have often seen risk assessment confused with risk management. Risk assessment is a component of risk management. Risk assessment is basically a quantification process that examines the hazards, impacts, and consequences of some situation that requires management. Risk management builds on the risk assessment process by addressing such questions as (1) What can be done? (2) What options are available and what are their associated trade-offs in terms of costs, benefits, and risks? and (3) What are the impacts of current management decisions on future options?

Similarly, programmatic risk management for long-term care should focus on broad issues associated with general management. It is a process of analyzing, selecting, implementing, and evaluating actions related to broad categories of risk. Managing risks is actually about managing an organization (or organizations) by planning, organizing, directing, and facilitating processes and resources to achieve the various objectives (Freudenburg, 1988). Effectively designing and implementing these management features are among the basic challenges associated with viable institutional management systems for long-term care of hazards.

Within this context, risk management and hence managing uncertainties involve many areas not traditionally considered in risk management approaches. In addition to the traditional budget, schedule, cost, environmental and worker safety issues, other areas include regulatory, cultural, social, political, legal, and quality issues. Because many of these topics are not readily quantifiable, risk management involves both quantitative and qualitative components.

Examples of some nontraditional risk issues include the consideration of cultural factors. With many of the large nuclear waste sites in the West located on or near Native American lands, the challenge for risk assessment is quantification of risk based on Native American lifestyles and not traditional urban and suburban residential scenarios (Harris & Harper, 2001). In the regulatory arena, regulations can change with time. Hazardous constituent maximum concentration limits may be lowered (or raised). Substances previously thought to be nontoxic may subsequently be determined to be toxic. A recent example of this is the chemical constituent perchlorate. Its impending regulation has sparked a heated national debate about acceptable concentration limits. What was once an unregulated substance will become a regulated hazardous constituent. Changes in acceptable concentration levels can vastly alter long-term care decisions for a waste site. Changes in social contexts over time are likely and should be anticipated. The challenge in risk management is understanding and incorporating those changes into the planning and implementation processes of long-term care. Individual and public values and needs can change quickly, and the long-term care system needs flexibility to accommodate those changes.

There are several risk based decision-making frameworks that have potential applicability in relation to long-term care issues and managing long-term uncertainties. Three examples include the Environmental Protection Agency's Integrated Decision Making Framework (EPA, 2000), the Government of Canada's report on risk management (2000), and the Presidential/Congressional Commission on Risk Assessment and Risk Management (1997). There are also some excellent European decision frameworks. These frameworks outline various steps in a risk-based decision-making process designed to address complex environmental issues such as climate change. The German Advisory Council on Global Change (1998) even proposed a decision framework that classifies risk based on Greek mythology. These diverse frameworks nevertheless tend to have similar components which include (1) problem definition, (2) analysis and decision-making, (3) implementation, and (4) evaluation and lessons learned.

The application of a programmatic decision-making framework for long-term care is something that should be considered for implementation by DOE and others to aid in the institutional management process. One of the many benefits of this approach is that these frameworks can help assist in transparent decision-making. It is particularly important that DOE (and others) utilize processes that allow for transparency and appropriate public input. DOE has long been criticized for its "decide-announce-defend" approach to policy development and planning. Since much of the burden of long-term care, particularly related to institutional controls, will rest with local communities and the states, it is incumbent upon DOE to utilize processes that have some consistency and involve the public throughout the process.

Another consideration in decision-making is reliance on technology. I have observed many instances where there is hope that new technologies will alleviate present-day problems. While new technologies are becoming available, the expectations for technology coming to the rescue have often proved unrealistic. However, investment in technology development should not be overlooked and should be pursued. Specific areas in need of development are treatment, monitoring, information management, institutional controls, and land use.

The need for technology development for basic science issues related to long-term care has been discussed by the National Research Council (NRC, 2000a, 2000b) and others (DOE, 2002). While there are many areas that could benefit from technology development, from my perspective the most needed and perhaps the most challenging are related to institutional controls and associated land use issues. My experience is that many people consider institutional controls to be readily available and implementable. I have often heard comments to the effect that "we'll just rely on institutional controls." A deed restriction by itself does not provide assurance that contamination will remain inaccessible. The expectation should be that institutional controls will eventually fail, as they have in the past (NRC, 2000a), leading to the need for further research and development. The Interstate Technology Regulatory Council (ITRC) recently surveyed a number of state regulators regarding their perspectives relating to long-term care (ITRC, 2004). One of their key conclusions was that the effectiveness of institutional controls can be enhanced by technology development.

Institutional Management Complexities

It has been my assertion that long-term care is and will be far more complicated and challenging than remediation activities. I believe this for several reasons, two of which are discussed below.

The first is that we have to consider how human systems work and can be managed in a long-term context. Examples of the complexities of managing human systems can be found in numerous safety programs including DOE's and also NASA's space program. These approaches rely on among other things redundant systems and layering. The erosion of an agency's culture of safety in the face of political and cost pressures was dramatically illustrated by events leading up to the space shuttle *Challenger* disaster (Vaughan, 1996). For the most part, DOE has not as yet considered or addressed the fallibilities of human systems in long-range strategic planning as is needed for long-term care.

Another observation is that many of the managers involved in environmental remediation and hence long-term care are typically scientists and engineers. These managers tend to seek to solve problems in a linear manner. It is my assertion that the complexities of long-term care do not lend themselves well to being solved in a linear context (cf. Travis, 2006).

Rhetoric of Certitude and Values

The concept of the rhetoric of certitude was previously discussed in this series by Rosa and Clark (1999). It is a concept that I have observed on many occasions. On one side, you have the scientists who are armed with their facts, figures and models. They feel they know the truth and are supported by their scientific methods and data. The "I know I am right" because of science and facts can lead to difficult encounters when dealing with members of the public and their perceived values however. On the other hand, I have observed members of the public also relying on the rhetoric of certitude when they "know" unequivocally that the scientists are lying. Both parties are capable of exhibiting the rhetoric of certitude, although typically at opposite ends of the spectrum.

Values play an important role in the rhetoric of certitude because often these opposing positions have their basis in widely differing values. As Rosa and Clark (1999) state, "Many of the judgments and conclusions of scientists, just as with laypersons, are often value-laden or motivated by objectives other than truth." Perhaps it is obvious that development of public policy relies on finding commonality of values between potentially opposing groups, but this is not always easy to accomplish.

I do believe that one of the many reasons the surface remediation portion of DOE's UMTRA Project was relatively successful was that the majority of the people involved in the process, including states, Tribes, and stakeholders, had common visions and values that supported accomplishing the cleanup, despite the desire sometimes to take different paths to the recognized objective. The UMTRA Project succeeded in remediating over 46,000,000 yd^3 of tailings and contaminated materials. The program involved 10 states and four Native American tribes and literally thousands of private citizens, and additional stakeholders.

Value conflicts seem to be one of the greatest challenges associated with environmental issues in general. Dietz (2001) and others have written of this. My personal experiences in dealing with the challenges of water in the West and endangered species and nuclear waste disposal issues all have revolved around value conflicts. The technical issues may be different, but value conflicts, which are not the only cause of conflict, are at the core of the challenges.

Public Trust

Public trust, as most people know, can take a long time to nurture and cultivate, and seemingly seconds to destroy (Slovic, 1991, 1993). Two recent examples with DOE come to mind.

The first is the DOE's recent exercise involving Risk-Based End States (RBES). This process was used by DOE to revisit various cleanup decisions and agreements across the DOE complex. There are many sites where this process was not well received. In fact, many local communities and state and federal regulators wrote letters of protest to the DOE about the concept and process. Part of the criticism of the RBES process was that DOE Headquarters imposed a short timeframe for the development of the site-specific plans and hence did not allow the public time for adequate input. DOE has stated repeatedly that the RBES process is not a decision-making process even though it has all the markings of just such a process. The Fernald site in Ohio, which reached cleanup agreements for the site in the early 1990s, is a prime example. The citizens were outraged that DOE now seemed to be suggesting that it would not honor these agreements that had been negotiated in seemingly good faith. The public trust of DOE has been significantly damaged by the recent RBES process.

Another example is the development of the draft Sandia Long-Term Environmental Stewardship Plan (DOE, 2001b). The process for the development of the plan involved substantial public input over several years. Sandia conducted a number of meetings and workshops to solicit public input. A draft was submitted to the local DOE for its review. The version of the plan that was released in August 2001 was substantially different from the public's expectations. At the core of the controversy was the very late change in DOE staff charged with review of the plan. The new DOE staff did not support the plan as it came to them and made significant changes in terms of stripping much of what the public had requested. Many members of the local community were upset and felt that DOE had betrayed them.

Both the RBES and the Sandia examples are instances where DOE lost significant public trust. In each case, the deeper problem seems to have been with what Todd LaPorte refers to as institutional constancy (LaPorte & Keller, 1996; LaPorte & Metlay, 1996). DOE's actions led the public to perceive that DOE was departing from its core cleanup mission, eroding public trust. That the two examples cited here took place over a time period of only a few years suggests that institutional constancy may prove to be a significant challenge for any agency involved in long-term care of hazards.

What Constitutes Failure

A key component of long-term care is various forms of monitoring. Asking what we are monitoring and why leads to the question of what constitutes failure.

For example, engineered barriers require performance monitoring (Kostelnik et al., 2006). In the case of UMTRA sites, groundwater monitoring is required at a defined point of compliance to aid in ensuring that the cover system is performing as required. Several UMTRA sites were remediated in place, including the Tuba City, Arizona site. The tailings were consolidated and capped with an engineered barrier. The surface remediation was completed prior to the initiation of groundwater cleanup activities that are expected to last decades. Monitoring at a prescribed point of compliance within a preexisting groundwater plume would not aid in determining failure of the engineered barrier, however, if the preexisting contamination is greater than the potential contamination from the failure of the cover system. This was the case at Tuba City. Lengthy and protracted negotiations with the regulator eventually allowed for deferral of point of compliance monitoring until groundwater remediation is deemed complete.

The determination of what constitutes failure and the potential approaches or remedies require discussions with a variety of parties during closure planning instead of waiting until there is a problem, either real or perceived. Not only does this help in contingency planning, it sets the stage for some up front agreements and a possible compromise with the regulators early in the planning process.

Implementation

Implementation of any long-term care program will be a significant challenge. When talking about public policy, I often like to paraphrase Mark Twain, who said that it was easy to quit smoking because he had done it a thousand times. It is relatively easy to write policy but the real challenges are getting policy implemented. An excellent article about the failures of the implementation of policies can be found in the article *Expertise and the Policy Cycle* (Barkenbus, 1998).

Regulatory implementation also has its challenges. Over the course of nine years, I assisted in negotiations of the final groundwater standards for the UMTRA Project with the EPA. There were many detailed discussions about the meanings of words such as "will," "should," and "could," as they have significant consequences in terms of the actual implementation of the final regulations.

An important component of successful implementation is the commitment of management to make it happen. In my experience, a large portion of the challenge lies with educating various layers of management about the intricacies of long-term care issues. I have heard more than once from management that they were addressing long-term care because they were going to conduct monitoring. In addition to the education needs, effective implementation of long-term care programs will require an ongoing, long-term commitment that few agencies and management have to date been able to demonstrate.

Corporate Knowledge

An additional factor in long-term care is the concept of corporate knowledge. While many recognize the need to document the remedial activities conducted at a site, an often-overlooked concept is the ability to understand past (historical) decisions in a contemporary context. Many remediation programs can last a decade or longer. The ability to understand how and why decisions were made 10 or 20 years earlier is often difficult when the answers are buried in reams of paper in remote storage. Various approaches to facilitate this knowledge should be considered (Drew, Nyerges, & Leschine, 2005). Narratives and timelines can aid in this. Sandia is currently utilizing this approach for dozens of completed sites.

The Two Colors of DOE Environmental Management

A current issue in the challenges of long-term care planning is the two different entities at DOE with environmental management responsibility. In 2002, the DOE created the National Nuclear Security Administration (NNSA). This branch, most of which was previously in the defense programs portions of DOE, is tasked primarily with weapons production and nuclear stockpile stewardship. There was also the Environmental Management portion of the same DOE, which was tasked with various environmental activities including remediation and long-term stewardship. Following DOE's 2006 reorganization, the Energy, Science and the Environment branch includes Environmental Management and Legacy Management, created in 2003.

In the case of Sandia, the majority of environmental remediation activities at legacy sites are due to be completed by the end of the 2006 federal fiscal year. At that time, the landlord, the NNSA, will assume responsibilities for long-term care of legacy sites at Sandia. This presents some challenges. As could be expected, one size fits all planning has some inherent problems when applied to two groups with different missions. Current discussions between

Environmental Management and NNSA at Sandia have centered on funding, resources, and trying to resolve different approaches to planning.

Of course, the majority of the public does not realize that there are two environmental management DOEs. Therefore, the public does not necessarily understand the challenges with the planning and funding and the two colors of DOE.

CONCLUSION

The issues associated with long-term care of contaminated sites are obviously very complex and challenging. Each issue has many components, layers, and contexts.

In the many years that I have been involved with DOE long-term care issues, it has been apparent that things have often not gone according to DOE plans. The track records for failure of human systems lead to the need to plan for and to a certain extent expect fallibility (NRC, 2000a). With this in mind, planning should be dynamic and flexible, and the various systems involved should be designed for layering and redundancy.

Long-term care planning has not been very robust or well implemented. Key reasons are the lack of understanding of the complexities, particularly by management; the need to manage for long-term uncertainties; the lack of clear regulatory drivers for long-lived hazards; the lack of adequate resources; and the seeming lack of long-term commitment by the Federal government.

As I stated previously, the DOE's UMTRA Project was successful for several reasons. These range from the foresight of the congressional legislation which specified the need to minimize active maintenance and requirement that the disposal cells be designed to be effective for 200–1,000 years. This led to the development of some redundant and layered components. An example is the requirement of the disposal cells to withstand probable maximum precipitation and flooding events. Additionally, the requirement that the federal government care for these sites in perpetuity lessens the primary concern of misuse. The UMTRA Project was also successful because Congress insisted on an active public participation process and the implementation of cooperative agreements with each state and Native American tribe involved in the program. These factors, along with the commitment of those involved to make it happen, led to the successful surface remediation of the sites. These varied reasons for the success of the

UMTRA remediations are some of the same components that many have recognized as key needs for any long-term care program.

It is the obligation of the Federal government to provide the requisite resources, focus, commitment, and regulatory drivers to protect public health and the environment from these long-term environmental hazards. The challenge is that the government has a history of lack of constancy, and long-term care in particular requires a constancy that has not been demonstrated. Moreover, there is a need for a vigilance that has not been previously demonstrated by bureaucracies. The Love Canal redevelopment is a much-cited example.

Public trust is often damaged by DOE, either knowingly or unknowingly. The recent risk-based end state process is an example. In fact, this is just another example in a series of frequent DOE mission changes that have taken place within the last few years. This, too, speaks to the lack of institutional constancy.

While there are significant needs for technology development in numerous areas, future technology development should not be expected to solve challenges of planning and implementing long-term care. A geochemist colleague of mine use to joke about the need to find the mineral "miraculite," as it would solve all of the geochemical problems for groundwater cleanups. In the same vein, a "miraculite" technology development should not be expected to handle all the future problems of long-term care.

Probably the most important challenge in long-term care is the issue of values. Changing values over time and value conflicts associated with long-term management of long-lived hazards will be difficult to predict, let alone manage. Frequent interaction with the various people and institutions involved in long-term management will be very important in order to access and incorporate value issues into planning.

It has often been said that one of the only things that is certain is change. Long-term care is about planning for and expecting change. It is all about expecting the unexpected.

REFERENCES

Barkenbus, J. (1998). *Expertise and the policy cycle*. Energy, Environment, and Resources Center, University of Tennessee, September.

Department of Energy (DOE). (2001a). *Long term stewardship study, Volume 1 –report. Final study*. Office of Long-Term Stewardship, Office of Environmental Management, U.S. Department of Energy, Washington, DC, October.

DOE. (2001b). Environmental Restoration Project Sandia National Laboratories/New Mexico in association with the U.S. Department of Energy, Kirtland Area Office Draft Long-Term Environmental Stewardship Plan. August.

DOE. (2002). *Draft long-term stewardship science and technology roadmap*. DOE/ID-10926, August.

Dietz, T. (2001). Thinking about environmental conflicts. In: L. Kadous (Ed.), *Celebrating Scholarship*. College of Arts and Sciences, George Mason University, Fairfax, Virginia, http://www.dietzkalof.org/dietz_research.html

Drew, C. H., Nyerges, T. L., & Leschine, T. M. (2005). Promoting transparency of long-term environmental decisions: The Hanford Decision Mapping System Pilot Project. *Risk Analysis, 24*(6), 1641–1664.

Environmental Protection Agency (EPA). (2000). *Toward integrated environmental decision making*. Environmental Protection Agency Science Advisory Board, Washington, DC, August 2000.

Freudenburg, W. R. (1988). Real risk, perceived risk: Social science and the art of probabilistic risk assessment. *Science, 242*(4875), 44–49.

German Advisory Council on Global Change. (1998). *World in transition, strategies for managing global environmental risks*. Annual Report, Bremerhaven, Germany.

Government of Canada, Privy Council Office. (2000). *Risk management for Canada and Canadians*. Report of the ADM Working Group on Risk Management, Ottawa, Canada, March.

Harris, S., & Harper, B. (2001). Lifestyles, diets, and native American exposure factors related to possible lead exposures and toxicity. *Environmental Research, 86*(2), 140–148.

Interstate Technology and Regulatory Council (ITRC). (2004). *Issues of long term stewardship: State regulators' perspectives*. RAD-3 ITRC Radionuclides Team, available at http://itrcweb.org.

Kostelnik, K., Clarke, J. H., Harbour, J. L., Sanchez, F., & Parker, F. L. (2006). A sustainable environmental protection system for the management of residual contaminants. In: T. M. Leschine (Ed.), *Research in social problems and public policy*. Amsterdam: Elsevier [... this volume].

LaPorte, T., & Keller, A. (1996). Assuring institutional constancy: Requisite for managing long-lived hazards. *Public Administration Review, 56*(6), 535–544.

LaPorte, T., & Metlay, D. (1996). Hazards and institutional trustworthiness: Facing a deficit of trust. *Public Administration Review, 56*(4), 341–346.

National Research Council (NRC). (2000a). *Long-term institutional management of U.S. Department of Energy legacy waste sites*. Committee on the Remediation of Buried and Tank Wastes, Board on Radioactive Waste Management. Washington, DC: National Academy Press.

NRC. (2000b). *Research needs in subsurface Science: U.S. Department of Energy's Environmental Management Science Program*. Committee on Subsurface Contamination at DOE Complex Sites. Washington, DC: National Academy Press.

Presidential/Congressional Commission on Risk Assessment and Risk Management. (1997). *Framework for environmental health risk management, final report* (Vol. 1). Presidential/Congressional Commission on Risk Assessment and Risk Management. Washington, DC.

Rosa, E. A., & Clark, D. L., Jr. (1999). Historical routes to technological gridlock: Nuclear technology as prototypical vehicle. *Research in Social Problems and Public Policy, 7*, 21–57.

Slovic, P. (1991). Perceived risk, trust, and politics of nuclear waste. *Science*, *254*, 1603–1607.

Slovic, P. (1993). Perceived risk, trust, and democracy. *Risk Analysis*, *13*(6), 675–682.

Travis, C. (2006). Vulnerabilities and uncertainties in long-term stewardship. In: T. M. Leschine (Ed.), *Research in social problems and public policy*. Amsterdam: Elsevier (... this volume).

Vaughan, D. (1996). *The Challenger launch decision: Risky technology, culture and deviance at NASA*. Chicago, IL: University of Chicago Press.

LONG-TERM STEWARDSHIP AND RISK MANAGEMENT: ANALYTIC AND POLICY CHALLENGES

Eugene A. Rosa

> To understand decision making in any organization, we must look at individual action within its layered context: individual, organization, and environment as a system of action. When all are taken into account, we find that they modify the notion of rational choice that undergirds the amoral calculator hypothesis.
>
> Diane Vaughan, *The Challenger Launch Decision* (1996)

On January 28, 1986 at 11:38 AM EST the Space Shuttle Challenger was launched amid great national fanfare from the Kennedy Space Center at Cape Canaveral, Florida. Seventy-three seconds after launch, O-ring seals were breached allowing the ignition flame to reach the liquid hydrogen and oxygen-fueling ascent into orbit. The result was a technological disaster. The Challenger erupted into a giant fireball, sending the crew compartment into the sea at 200 mph. All seven crewmembers died, among them science teacher Christa McAuliffe, the first citizen-passenger launched into space. The cause of the disaster – heightened in drama by its live television coverage to a stunned nation – could hardly be more straightforward: it was a technological failure. The O-rings that were supposed to prevent the ignition flame from reaching the liquid fuels had not done their job. But, why not?

The question begs a much deeper set of questions: How could an organization comprising the most elite group of scientists and technologists known to the world, NASA, and routinely challenged to manage complex

Long-Term Management of Contaminated Sites
Research in Social Problems and Public Policy, Volume 13, 227–255

ISSN: 0196-1152/doi:10.1016/S0196-1152(06)13010-0

technologies, have failed? What lessons can be learned from this organizational disaster? And what is the appropriate analytic frame for understanding technological failures such as this? For only by choosing a proper orientation or conceptual template can we hope to understand not only why the O-ring failure on the Challenger led to disaster, but also why failed management can result in a variety of large-scale technological disasters. It is these deeper questions that are our focus here. The long-term stewardship of not technologies themselves, but the wastes produced by technologies, is the context for examining these questions. Long-term stewardship is a complex, newly recognized management challenge.

LONG-TERM STEWARDSHIP

Long-term stewardship (LTS), the caretaking of hazardous materials, is one of the main unanticipated challenges of high modernity. LTS refers to the process of protecting public health and the environment through the effective management of systems or materials over multiple generations, in some cases over many many generations. It arises from the recent realization that the full remediation of contaminated waste sites is beyond scientific knowledge, best technologies, or available resources.[1] Some materials will demand care and risk management over several generations while others, such as high-level nuclear waste, will require a succession of generations that exceeds the longevity of any civilization known to history.

In the United States there are hundreds of thousands of sites with residual radioactive or chemical contamination that are a risk to the health of citizens and the environment (Pastina, 2004). The stewardship of these many and varied sites rests in the hands of major departments and agencies of the federal government: Department of Energy (DOE), Department of Defense (DOD), Department of the Interior (DOI), Environmental Protection Agency (EPA), National Aeronautics and Space Administration (NASA)[2] as well as the states numerous municipalities, and Indian Tribes. While there are successful examples of stewardship, a group convened by the National Academy of Sciences concluded that "there is little evidence that lessons have been learned from cases of successful stewardship elsewhere in the Federal estate …" (Pastina, 2004).

LTS and Risk

In all cases, LTS comprises systems and materials that have the potential for catastrophe, for environmental contamination, or for inflicting injury, ill health, or death on exposed humans. Thus, one foundational component of virtually any LTS system will be risk assessment, the active anticipation of the possibility of untoward outcomes. Indeed, risk has been so identified in efforts by the National Academy of Sciences Steering Committee on Long-Term Stewardship to develop a National Academy study program designed to map the challenges and the parameters of a successful LTS program for the broad spectrum of wastes (Pastina, 2004).

Because, as noted above, few successful stewardship programs currently exist, and because it is not clear what lessons are to be drawn from our relatively limited experience with stewardship systems to date, models are not available for guiding the development of stewardship strategies. Hence, the development of those strategies must rely on analytic frameworks and risk management experiences from other contexts. One obvious source is the risk assessment and management literature comprising, as it does, three decades of cumulative knowledge.

There are three limitations inherent in the risk literature, however. First, many of the risk management efforts of stewards will extend far into the future, with literally no precedents as a guide. The longer the extension into the future the less certain are estimates of both the probabilities and consequences (the two defining features in the standard definition of risk) of risks until they are realized – if ever they are. Over long-enough periods of time our knowledge of the risks is effectively zero.

Second, the risk literature is overwhelmingly dominated by the rational actor paradigm (RAP), an entirely reductionistic, analytic frame for assessing risks and decisions about them. RAP's focus on individuals and its meager purchase on the decisions of organizations are key limitations that will be addressed later. LTS, whatever the particular management task – short or long – will rest on the shoulders of organizations and institutions, not individual actors.

Third, in pure form RAP treats the individual as an abstract, calculating decision-maker unmindful of social and political concerns and committed to making reasoned judgments. A presupposition of RAP is that optimal choices will apply to not only the present abstract decision-maker, but also to abstract past and future decision-makers. Such a presupposition requires the suspension of knowledge about history, on the one hand, and a naiveté about predicting future preferences, on the other. By definition LTS is a management process extending over time. Hence, it requires an agent that promises continuity over the required time of stewardship. Only organizations and institutions, not individuals, can fulfill this role. The limitations of RAP for such management challenges are clearly revealed in analyses of the Challenger disaster.

RAP and the Challenger Disaster

In early February 1986, then President Reagan created a commission to investigate the Challenger disaster. Headed by former Attorney General William P. Rogers, this Presidential Commission (also called the "Rogers Commission") identified the O-rings as the cause of the tragedy. But the Commission went far beyond this proximate causal connection and investigated the circumstances that gave rise to this technological failure. A bi-partisan Congressional Commission performed a similar investigation at about the same time.

The reports of both bodies, despite several points of disagreement, were similar in outlook and in the attribution of the accident's cause: it was due to wrongdoing by NASA personnel.[3] "First and paramount, the Presidential Commission indicated that NASA personnel had violated both industry rules and internal NASA rules designed to assure safety" (Vaughan, 1996, p. 33). While both commissions recognized the importance of the successively larger contexts that shaped these rule violations, and while both concluded that these contexts facilitated wrongdoing, in the final analysis it was individual decision-makers who were to blame.

The general explanation to crystallize from commission reports and media coverage elaborated a trickle-down logic by specifying a set of causal conditions leading to the accident. Political pressures to demonstrate the safety of the shuttle program (i.e., it was now safe enough to invite civilians aboard) combined with production pressures (namely compliance with launch schedules necessary to ensure profitability of a self-sustaining shuttle program) shaped NASA administrators' actions. This resulted in decisions at lower levels involving rule violations and individual wrongdoing. Cost-benefit calculations were shaped by these conditions, resulting in decisions that sacrificed safety for the sake of political and economic gains. The decisions of managers to knowingly violate safety requirements, while rational, were amoral.

At the bottom of the Commissions' Reports claiming that rule violations caused the O-ring failure and, therefore, the disaster, lies a theoretical orientation: the Rational Actor Paradigm. RAP assumes that human beings are cognitively equipped, purposive actors seeking to make choices that are subjectively best or optimum in some way. The decision to launch the Challenger – indeed, all space shuttles up till then – was a decision based upon rational cost-benefit calculations. Because of the fatal outcome of Challenger, the rational decision was judged and named after the fact as "amoral" (Vaughan, 1996). By this view, managers become "amoral calculators" who violate regulations if need be to assure organizational goals are attained: "When decision makers' calculations of costs and benefits are tainted by self-interest, economics, or politics so that *intentional* [emphasis in original] wrongdoing and/or harm result, their calculation becomes amoral" (Vaughan, 1996, p. 35).

This amoral calculator hypothesis is not unique to the Challenger disaster. In fact it is a leading explanation of wrongdoing by organizations generally, despite the fact that there is remarkably little empirical support for it. Indeed, the hypothesis has been seldom tested (Vaughan, 1996). Its theoretical integrity rests upon a fundamental presupposition that managers within organizations are relatively free to make rational cost-benefit calculations. Yet it is unlikely that organizations would survive under the resulting conditions of decision-making anarchy. Instead, all organizations reduce the uncertainty of individual actions by imposing expectations, routines, and rules of choice and action. Decision-making is, therefore, constrained by premises of permissible choice embedded in the structure and culture of organizations. This reasoning reaffirms a fundamental orienting principle in sociology – namely that context matters.

It is from this contextual orientation that Diane Vaughan (1996), in the most comprehensive academic analysis of the accident to date, was able to demonstrate the logical flaws and problematic conclusions of the received view of the disaster. Rather than amoral but rational calculations, rather than good people doing dirty work, rather than individual wrongdoing – Vaughan finds the cause of the accident in the very structure and especially the culture of NASA itself. The rules of that culture included bureaucratic procedures that permitted the acceptance of engineering performance that did not meet design standards, a structural secrecy where the left hand was out of communication with the right, and the acceptance of the risk of serious failure based, not on positivistic foundations, but on a cultural construction. In short, people were doing their jobs exactly as prescribed by the NASA bureaucracy and culture. And it was this "rightdoing," not wrongdoing as concluded by the two Federal reports, which caused the accident.

On the one hand, the theoretical and management insights provided by Vaughan's analysis of the Challenger disaster is hidden from RAP. On the other, the analysis points to the importance of organizational analysis of technological failures – especially where the technology failure reflects, as it does in this instance, an organizational failure. Hence, there is the clear need to broaden the range of analytical approaches to risk analysis, especially for risk choices in the hands of organizational and institutional actors – such as

LTS. Arguing for organizational analysis means to abandon neither RAP nor the psychometric and other individual-centric approaches that overwhelmingly dominate the risk field, but to complement them with other, more holistic approaches to risk. In particular, RAP should be augmented with organizational analysis. The promise of an organizational approach is demonstrated with a review of the key areas of research in this incipient risk paradigm. (But first, we critically review the dominant RAP.)

THE RATIONAL ACTOR PARADIGM IN THE STUDY OF RISK

The modern field of risk analysis is dominated by RAP, a theoretical orientation with roots in past centuries.[4] Its distinguishing characteristics are its assumptions about human behavior and its theoretical methodology.[5] RAP assumes that abstract human beings are cognitively equipped, strategic, goal-serving actors who make choices that are subjectively best or optimum in some way. RAP's theoretical method is a reductionistic "methodological individualism," an approach where the term "individual" is used in the extended sense to refer to not just the individual decision-maker, but also to any individual entity: group, firm, organization, government.

Formalizations of risk analysis unmistakably reflect RAP's underpinnings: actuarial analysis, cost-benefit analysis, probabilistic risk assessment (PRA) (Rasmussen, 1981) and revealed preference (first appearing in the pioneering article by Chauncey Starr (1969)). Despite considerable criticism (Slovic, Fischhoff, & Lichtenstein, 1977; Kahneman & Tversky, 1979; Tversky & Kahneman, 1981), RAP continues as the reigning perspective in the field. This continued reign has resulted in an inattention to alternative perspectives for understanding risk, especially perspectives – such as organizational analysis – whose theoretical method is not the methodological individualism of RAP, but methodological holism. With methodological holism the proper unit of analysis is not the individual (however extended), but some aggregated unit, such as a formal organization, appropriate to the research questions framed. Hence, the appropriate strategy for understanding the risks of LTS is to examine the organizations that are charged with stewardship.

RAP and the Inevitability of "Operator Errors"

Because RAP builds its edifices from a foundation of methodological individualism, it directs attention toward individuals and away from aggregate units with emergent properties like organizations. This is also consistent with the more general tendency – among laypersons and experts alike – to view social phenomena in individualized, human terms ("anybody can become president"). This micro-bias, a layperson form of psychological reductionism, is manifest throughout the risk field. For example, as Perrow (1986, 1999) has pointed out among high-risk organizations approximately 60–80 percent of all accidents are attributed to operators when the correct figure is probably about half that. Historian Edward Tenner (1996) not only echoes this observation, but also generalizes it by observing that "The better and the safer technology becomes, the more we presume human error when something goes seriously wrong" (p. 261).

The same micro-bias emerged with the release of the Kemeny Report (President's Commission on the Accident at Three Mile Island, 1979) after the Three Mile nuclear accident in Pennsylvania in March 1979. The key conclusion of the much-publicized report, at least as it was interpreted by vested interests, the press, and other interested parties was that the accident was due to operator error.[6] This practice of blaming individuals is a logical consequence of reductionistic, RAP-based approaches such as PRA, where hardware and humans are treated alike – as system components. So, when the system fails, it is logical to look for the cause in components, especially human ones over which accountability can be imposed. This propensity among experts and laypersons alike to seek failure causes in individual actions may be described as the "sociological attribution error" – a mindset that attributes blame to individuals as a consequence of inattention to the organizational structures and cultures that drive individual choices.

Organizations as Key Risk Actors

Organizations are a ubiquitous feature of modern life and one of its most powerful forces. They dominate the landscape of not only social collectivities but also the daily experience of a large majority of citizens in modern societies. As Charles Perrow notes, "all social processes either have their origin in formal organizations or are strongly mediated by them" (Perrow, 1986, p. vii). For many observers, modern societies are "organization societies."

The role of organizations is pivotal in the study of risk. The introduction, management, and control of technology are overwhelmingly in the hands of organizations. The importance of organizations – their structure, culture, and operations – to the management of technological risks is clearly illustrated in the Challenger disaster. Other major technological disasters – such as the Three Mile Island nuclear accident, the chemical accident in Bhopal, India, and the nuclear accident at Chernobyl, Ukraine – can likewise be traced to organizational mismanagement or organizational failure. These four disasters are easily recalled when thinking about numerous other risky technologies whose management is the responsibility of organizations of widely varying form and purpose. Yet despite this reality, the field of risk remains dominated by reductionistic approaches that ignore organizations as structures by focusing on individual components, individual decision-making, or by treating organizations as if they were aggregated individuals. On what theoretical base does this practice rest?

RAP and Organizational Analysis

The idea that RAP is a fruitful theoretical organization for understanding organizations enjoys a long history in the field of organizational analysis. The "decision-making" school of organizations, especially the work of Simon and March (March & Simon, 1958; Simon, 1957; March, 1978), strove to answer the question: Do organizational actors, real individual or legally constructed entities, engage in rational action? More specifically, does RAP shape their criteria of action, and does RAP account for their behavior? Simon and March addressed the rational actor question not via assumption, but with a clearly developed analytic strategy. They intended to explain both the actions of employees making up organizations and those of the organizations as an entity. An understanding of the former, the individual actions, would be used as a basis for understanding the latter, the organizational ones.

With individuals as the unit of analysis, the strategy was to build upon that foundation until the organizational edifice itself appeared. Logical and empirical difficulties conspired against the theory, however, and the presumed goal of integrated coherence was never realized. Nevertheless, this theoretical tradition left us with a variety of insights – many of them now classic "gems" that are accepted contingencies or conditionals in the application of RAP.

Humans have every intention to act rationally. We have a wide variety of tastes, desires, and goals that require the evaluation of alternative means for satisfaction. With a clear vision of goals, and with thoughtful assessment of the best means for achieving those goals, we are seen to be capable of rational decisions. But, despite sincere intentions to act rationally, we often do not. Cognitive limitations in our powers to reason rationally are one of the reasons. Another is that there are often unacceptable costs, such as time and money associated with the search for alternatives, requisite with the attainment of the one best rational outcome.

A rational decision is often defined as the one resulting in an optimal outcome, where optimal refers to the most preferred of all available alternatives. Often the criterion of "most preferred" is the alternative that maximizes some commodity, such as money. Frequently, however, we do not seek and choose the optimal solution, but one that we consider satisfactory; we do not optimize, rather we "satisfice." For March and Simon (1958), "an example is the difference between searching a haystack to find the *sharpest* needle in it and searching the haystack to find a needle sharp enough to sew with it" (p. 141). We try to act fully rationally and sometimes do, but we most often engage in a suboptimal form of rationality: bounded rationality.

With the discovery of our bounded rationality, how can the rational actions of organizations be explained within RAP? First, by ignoring the limitations of individuals via simplifying assumptions, thus abstracting away the complexity of the individual with a shift in the unit of analysis to the organization. Second, by removing the actor's volition to engage in satisficing or other suboptimal decisions by circumscribing her decisions, or by shaping the premises of the individual's decisions. In the second instance, the decision focus was shifted from individuals to the premises the organization imposes upon individuals intending to make rational decisions. These premises flow from the conjunction of the principal features of classic bureaucracy: division of labor, routinization, universal standards of performance, an authority structure, communication channels, and indoctrination – both procedural and ideological.

The upshot was that the organization was seen to be a rational entity seeking to perform rational decision-making – a *persona ficta*,[7] a fictive, but active rational actor, not simply a hierarchical authority structure resulting from rationalization of social processes. Furthermore, the propensities, capabilities, and limitations of individuals in the organization were explicitly recognized, thereby deepening the analytic framework and signaling dangers in organizational theory construction that begin with individual action. By explicitly recognizing the individual limitations posed by bounded rationality, the approach revealed more directly a chief function of the organizational form: to impose, top-down, rational constraints on the suboptimizing individuals so that rational outcomes could be generated. The organization could achieve optimal solutions even if individuals did not search out optimum alternatives. Perrow (1986) aptly put it this way: We can see "... in the organization a means of controlling the individuals in the interests of the goals of the leaders of the organization. The organization is more rational than the individuals because order is imposed on members by those who control the organization, and the order is in the interests of goals and purposes established and guarded by those in charge" (p. 68).

Thus, an unadulterated RAP had to be abandoned as a basis for Simon and March's first objective: to demonstrate that individuals are rational actors who obey the provisions of RAP. For the second objective – the demonstration that organizations are also rational actors – RAP served its purposes well. Organizations could be analyzed and understood as rational actors seeking optimal solutions.[8] But even here the orientation of RAP eventually bumped up against limitations in explaining the risk actions of organizations. One of these limitations is its inattention to the structure of information and power in organizations. To all organizations flows information. And within all organizations flows information. Information is the raw material for the rational estimation or reconstruction of risk decisions. What or who is at risk is often dependent on the type of information available, how it is amplified or otherwise massaged, how it is edited in transmission among actors, and the use to which it is ultimately put.

RAP assumes that the perfectly rational actor will collect all the necessary information necessary to delineate all alternatives and then choose the one that meets maximization or minimization criterion, as the case may dictate. In the complex world of conflicting goals, resource and time constraints, and uncertain external environments, managers often eschew formal analyses and take short cuts. For example, top-level executives often prefer "face-to-face interaction and even hearsay and gossip to analysis of factual reports" (Mintzberg, 1973, p. 118). And in the contextualized world of organizations the process of risk assessment is often reversed. Because they are hierarchically structured, organizations have elites sitting on top who exercise considerable control over information. Often their use of information is not to be rational, but to rationalize – that is, to justify actions already taken and to reaffirm their authority. The consequence is to make risk management strategies the dependent variable, not the independent variable assumed by RAP.

Indeed, Clarke (1989) and others have convincingly demonstrated that formal risk analysis – an operational tool of RAP – is often applied, not to establish quantitative estimates of risk, its intended purpose, but to rationalize management decisions already in place. The formalities of risk analysis are often only afterthoughts to political, legal, and economic considerations.

SOCIOLOGY OF RISK: TURNING THE MACRO LENS OF ORGANIZATIONS TOWARD RISK

Normal Accident Framework

The most direct attempt to understand risk from an organizational perspective was undertaken two decades ago by long time organizational theorist Charles Perrow. Perrow (1984, 1999) turned his analytic attention to risky technologies, with the principal product of his efforts the now classic book, *Normal Accidents: Living with High Risk Technologies*. It is a provocative book with a provocative thesis: that some of the complex technological systems we create, because they are beyond our complete understanding and management, are inherently risky – indeed, they cannot be made safe. Human technological reach exceeds organizational grasp. Of deep concern is the fact that a growing number of technological systems are highly risky, prone to failure, and have the potential for catastrophe. Indeed, for certain types of systems the organizations managing them are configured to produce not just the rare accident, but "system accidents" that are inevitable or "normal." Thus, to the central risk question – Can all hazardous systems be made entirely safe? – the answer is a highly pessimistic "no."

The scope of Perrow's investigation is technological systems where a major failure can produce catastrophic consequences. The raw materials for his finished product, a new theoretical framework for thinking about high-risk systems, are nuclear reactor failures, airplane crashes, accidents at sea, dam failures, space missions, recombinant DNA technology, and nuclear weapons. All of these types of systems have had catastrophic accidents or near-accidents. Since these systems are large and largely pervasive, it is doubtful that any individual is free of the risks they impose. This broadens the range of possible accident victims, from first-party victims (those directly operating a technology) and second-party victims (support staff or users of a technology) to third-party (innocent bystanders) and even fourth-party victims (fetuses and future generations). That broadening is particularly apt for LTS since a key LTS risk is to third and fourth parties.

Normal accident framework (NAF) emphasizes the political nature of organizational life: organizations are not rational entities, but entities that comprise actors with competing individual and group interests. Owing to these political realities, these entities are "organized anarchies." Within this broad "open natural systems" perspective Perrow adopts garbage can theory, developed by Jim March and his colleagues (Cohen, March, & Olsen,

1972) as the conceptual tool for understanding these organized anarchies. Garbage can theory emphasizes bounded rationality (March & Simon, 1958; March, 1978), unexpected interactions, and competing group interests within an organization. Organizations are viewed as large political garbage cans that collect solutions looking for problems to attack, that collect problems that are typically "ill-defined," and where politically motivated actors promote their self-interested solutions to the ill-defined problems.

Perrow's approach is structural because it is the system itself that is the unit of analysis. The importance of NAF to an understanding of risk cannot be overemphasized, for it represents the first pure structural approach to risk. It is also one of the first detailed risk analyses that relies on methodological holism rather than methodological individualism. The structural approach and methodology represent a direct assault on the narrowness of RAP – especially the highly reductionistic version of RAP underlying the engineering systems approach to risk.

Engineering, the dominant perspective for estimating risks of technological systems, views such systems as a collection of components – including "human" components, as we have already noted. To understand the entire system, then, is to understand each of its components, their functions, and their interactions. Component failure is unavoidable and expected, so safety systems are included in system design. Safety is imposed on the system principally by designing in buffers, redundant components and backup systems, all ready to spring into action when a technological failure begins to occur. Human vigilance and performance is assured through ergonomics, where attention is directed to the design of those system components, such as control devices, with which humans interact. Safety is accomplished with better bells and whistles – literally. Most safety attention is directed to dials, warning lights, switches, and audible alarms – their design and placement. The overall risk of the system is assessed within the same engineering framework, with the reductionistic, componential approach of PRA.

RAP first enters complex sociotechnical systems via the engineering approach to risk and safety described above – that is, via risk assessment. It makes a second entry via risk evaluation and acceptance, in Perrow's (1986) words, through "... a rational, calculative marketplace theory of cost-benefit analysis" (p. 309). Economists and other risk assessors convert risks to a common metric, typically body counts or dollar counts, so that costs and benefits can be compared among competing risks. Less costly risks are to be preferred over more costly ones. One dollar is equivalent to any other, as is any body equivalent to any other (Ackerman & Heinzerling, 2004). A vexing omission of this approach, from a sociological perspective, is its absolute inattention to the distribution of risks and benefits. Are those who bear the risks of technologies also the ones who enjoy their fruits?

According to Perrow, the dominant, reductionistic approach to estimating the risks of complex technologies is seriously flawed. Component failure is inevitable and expected by engineers. Safety is designed with this in mind. But system components can interact in unexpected and unpredictable ways. Two or more small component failures often combine in unimagined ways to produce failures in the entire system – "system accidents." This is not supposed to happen because technological systems have many built-in safety features: redundancies, backup systems, control devices, and procedures of vigilance. Yet, small multiple failures can defeat (and have) the most elaborate safety systems. Because multiple failures are unexpected, they are not visible to system designers and are, therefore, outside a conscious purview of design or control.

A fuller understanding of complex technological systems requires not just the dominant reductionistic engineering analysis, but also a more macroscopic perspective.[9] Perrow's central analytic device is a typology of technological systems. Systems are categorized along two independent dimensions: one that ranges from linear to complex and a second that ranges from loose to tight coupling. Cross-categorized, the two dimensions produce a four-cell typology where each technological system receives its proper analytical placement.

The first dimension, linear-complexity, refers to system interactions. Key variables are the degree of specialization, spatial proximity of components or tasks, separation of dependencies, common mode connections, and feedback loops. Linear systems have high degrees of specialization, spatial separation, and few interdependencies, none or few common mode connections, and none or few feedback loops; complex systems have all these things. A linear system (the assembly line is its prototype) is not necessarily laid out in a straight line from operation A to operation Z, but one where the sequences are straight between any series of operations and are predictable.

The second dimension, loose-tight coupling, refers to the way interactions mesh: how deeply production processes are time-dependent, invariantly sequenced, flexible in reaching production goals, and sensitive to the degree of slack in the system. Tightly coupled systems are time-dependent with rigid schedules, sequence invariant, production task invariant, composed of many redundancies and buffers, and with little slack; loosely coupled systems are generally devoid of these characteristics. And coupling is the key to recovery from initial failures. Loosely coupled systems are better able to recover because there is more time to devise solutions; there is more flexibility in the substitution of supplies, equipment or personnel; and there are alternative ways of doing things. In short, there is more opportunity to seize upon proximate and emergent solutions.

The most catastrophe-prone systems are those, like nuclear power and nuclear weapons, that are interactively complex and tightly coupled. This convergence of complexity and tight coupling of certain socio-technical systems leads to an ineluctable organizational dilemma. The organization styles necessary to meet the two functional needs, complexity and tight coupling, of the riskiest socio-technical systems are inherently contradictory. Tight coupling requires central control for efficiency. Centralized systems, hierarchically arranged, are needed to ensure smooth production processes and effective compliance to emergencies that may arise. They are especially suited to routine tasks. Interactive complexity requires decentralization, with a provision for local autonomy. Decentralized systems, flexibly arranged, are needed to foster imagination and creativity to solve nonroutine, unexpected problems. Decentralization, because it allows operators to intervene when inevitable errors occur, is more forgiving and more effective in recovering from an initial failure. Centralized systems tend to be brittle, decentralized ones tend to be pliable.

Hence, a first step in devising effective management strategies for LTS could be to delineate the organizational dimensions, complexity, and linearity, of the vast array of types of waste sites. A second step would be to match up the conjunction of these dimensions with appropriate management structures. And a third step would be to rank the sites on their levels of risk and management challenge in order to determine those demanding the most management attention.

It is difficult, if not impossible, to devise organizations that are centralized and decentralized at the same time. The result is an intractable organizational dilemma for systems that are both interactively complex and tightly coupled. The two organizational styles are, therefore, unavoidably contradictory and, says Perrow, the twain shall never meet. The consequence? unavoidable or "normal" accidents. For example, the paradigm high-risk technology for Perrow is nuclear power. It is a complex, tightly coupled system, the very type of system prone to unexpected interactions. Indeed, that is apparently what occurred in the Three Mile Island nuclear accident in the United States (state of Pennsylvania) in 1979; four trivial failures interacted in mysterious ways to produce a "system" accident (Perrow, 1986, p. 151; 1999). The theoretical logic of the organizational dilemma seemed compelling enough to permit Perrow (1984, p. 348; 1999) an unconditional, singular forecast: "I would expect a worse accident than TMI in ten years – one that will kill and contaminate." The chilling forecast hardly needed to wait a decade, but only seven years when in April 1986 the Chernobyl nuclear accident did kill and did contaminate – much of Europe and elsewhere.[10]

Insofar as Perrow's central finding, that some technologies cannot be made safe, is valid, it raises some deeply disturbing questions. The questions are especially relevant for societies deeply textured with complex technologies, for others seeking a future with a similar social landscape, and for the risky residuals of those technologies that require LTS. If certain complex technological systems cannot be made safe – an unavoidable social cost – is this cost worth it? Do their benefits outweigh the widespread dread that many people feel for them? Does their operation reaffirm or redistribute the "life chances" among members of society? Do they entail overt political costs – such as favored legislation, intellectual capital, or subsidies by the state – or covert political costs – such as a loss of personal freedom so that appropriate security systems can operate? How does the NAF inform LTS? These and similar questions beg of additional sociological efforts to understand the structural features of risky technologies.

High-Reliability Organization Theory

"Working in Practice but not in Theory" is the arresting title of one article coming out of work by a group of mostly University of California at Berkeley social scientists (LaPorte & Consolini, 1991). It captures the essence of a completely different approach, a "closed rational systems" (Scott, 1992) approach, to understanding organizations managing complex risks and hazards (Morone & Woodhouse, 1986; Wildavsky, 1988; and the Berkeley group). And it captures a completely opposite set of conclusions from Perrow's: that serious accidents with hazardous technologies are not inevitable, but with proper design and management are preventable. In a coordinated suite of studies, over a period of several years, (see, e.g., LaPorte, 1988; LaPorte & Consolini, 1991; Roberts, 1993) the Berkeley group produced the most sustained body of empirical work to support this approach. That work focused on organizations whose operations are technologically complex and inherently risky, often with catastrophic potential. Carefully examined through extensive field work were three hazardous organizations: the Federal Aviation Administration's (FAA) air-traffic control system, the Pacific Gas and Electric Company's electric power system (PGE was at that time one of the largest utilities in the country and included in its system is the

Diablo Canyon nuclear power plant), and the peacetime flight operations of two U.S. Navy aircraft carriers.

These organizations perform with such remarkable smoothness and safety they have been named HROs. The Berkeley group offers an explicit definition of HROs:

> Within the set of hazardous organizations there is a subset that has enjoyed a record of high safety over long periods of time. One can identify this subset by answering the question: "How many times could this organization have failed resulting in catastrophic consequences that it did not?" If the answer is on the order of tens of thousands of times the organization is "high reliability" (Roberts, 1990, p. 160)

The findings are remarkable because they fly in the face of beliefs, as exemplified in the work of Perrow, about technological systems: that high-risk technological systems are inherently dangerous and human actions cannot remove the danger entirely. To these pessimistic conclusions, the HRO theorists reply with an equally optimistic set of conclusions.

An examination of HRO safety records reveals near-flawless operations. They are shining examples of organizations that know how to properly manage risky technological systems. But how? The answer seems to lie in bureaucratic structures that are hierarchical and rigid during routine operations (the majority of time), but flat and flexible during times of exigency or crisis. HROs, thus, apparently have "switch-hitting" organizational structures, shifting to either of two modes as conditions demand. Yet, the results have yet to find their place in an integrated theory. Instead, we are left with rational performance absent a specification of the presumably rational structure producing it – a set of complex, but orderly findings in search of theoretical explanation.

While the HRO framework has not yet led to an integrated theory, it did deliver the raw materials for theory building. It has identified four key strategies and processes found in organizations enjoying model safety records:

- the making of safety and reliability a priority by the organization's top leadership,
- designs that include high levels of redundancy, technical and personnel, in safety systems,
- the cultivation of a "high-reliability culture" that enacts itself in decentralized and continually practiced operations, and
- organizational learning through trial and error.

First and foremost among the requirements of a highly reliable system is an unwavering commitment to safety and reliability, from the very top of the organization. Other organizational goals, such as efficiency, organizational prestige or profit making, must continuously be subordinated to avoiding serious operational failures. Such a commitment also lays the groundwork for the third feature of HROs: the development of a high-reliability culture.

HRO researchers assume, as do normal accident proponents like Perrow, that human beings are boundedly rational; we are not the perfectly rational automatons of overly rational theories and fiction, but social actors with bounded objectives and limited cognitive capacities. But, unlike normal accident proponents, HRO researchers assume these limitations can be overcome by making redundancy and overlap key design features of organizations. Organizations can safely manage hazardous technologies through duplication (two different components performing the same function), through technical redundancy (such as backup electrical systems, backup computers, backup safety devices), and even through personnel redundancy (such as multiple personnel capable of performing the same job). Safety is also enhanced when an organization ensures considerable overlap between technical units (such as the use of multiple forms of communication that complement each other) and between personnel units (such as where personnel are given overlapping responsibilities). As one theorist put it, "duplication is a substitute for imperfect parts."

Redundancy is the designed way of responding to component failure, but safety is doubtlessly enhanced if components do not fail in the first place. A reduction in component failure improves chances that redundant systems will not be taxed beyond their capacity to respond. The principal management strategies for reducing the burden on redundancy are decentralization, a "culture of reliability," and the prevention of complacent routinization. Decentralized authority is necessary to safety because in time of crisis or emergency, there is often too little time to go up the hierarchy before responding. Safety-threatening situations require rapid and appropriate responses by personnel proximate to the situation with the requisite knowledge and skills. While decentralization is the necessary structural condition of safety, the sufficient condition is an appropriate culture. Organizations must recruit, train, and socialize personnel to have a shared belief and commitment to safety and to assume their decentralized authority in time of crisis. And to ensure that this culture does not become complacent – with the routinization of safety activities and the perfunctory performance of duties – there must be continuous on-the-job practice simulations that are realistic, challenging, and the basis for improvements in procedures.

The final requirement of HROs is the willingness by the organization to learn. HROs emphasize the importance of trial-and-error organizational learning in the management of risky systems. Procedures and management practices, however sound, are never sound indefinitely. As conditions and circumstances change procedures and routines must be adjusted and modified – especially where the changed conditions can impact safety. These adjustments and modifications are accomplished through trial-and-error, often in response to past safety incidents. But HROs often engage in anticipatory trial-and-error as well, performing thought experiments, simulations, and exercises in order to search for errors before they actually happen.

To the question of whether accidents (especially catastrophic accidents) are preventable, or are an unavoidable feature of risky, complex technologies we are left, not with a categorical answer, but with an engaging puzzle. HRO theory optimistically answers the question in the affirmative, that we can develop and operate hazardous systems safely. But it leaves unaddressed the unique challenge to LTS, namely whether any organization or institution can develop a safety culture with staying power? Can we devise organizations whose continuity is congruent with the time horizons of stewardship responsibilities? The NAF gainsays the optimism of HRO with a pessimistic answer. Which is correct? At present we have little basis on which to decide. The question is, therefore, ripe for investigation that systematically subjects the competing theories to empirical test.[11]

Organizations as Avatars of Murphy's Law

As noted above our understanding of risks could be deepened significantly if we recognize that organizations are major actors, not only in assessing and managing risks but also in the very creation of risks. An important starting point in this evolving perspective is the recognition that organizations, even the ones operating technologies or regulating their safety, can "amplify" real risks. Often, they actually do. Embedded within this orienting premise is a tacit, but forceful criticism of RAP as an explanatory framework for organizational behavior. For despite their supposed rational systems, organizations produce subrational outcomes: amplified risks. The organizational factors responsible for amplification are varied. One factor derives from a fundamental challenge of organizational life: how to ensure that risk management and safety come first. These goals are often of only secondary or tertiary concern to an organization, a necessary management evil to be dealt with in the course of the organizations core goals, such as profit making.

"Organizations seek to produce energy 'while' protecting the environment, operate submarines while providing an adequate level of protection for the crew, dispose of wastes 'in an environmentally acceptable manner,' and so forth. Almost never is risk management included in the first half of the sentence, at least in the description of overall organizational goals, as in 'increasing the level of safety for workers in nearby communities while maintaining adequate profit margins'" (Freudenburg, 1992, pp. 21–22). Concerns with safety are often viewed as diversions – diversions of management attention and resources to peripheral activities. In time of shrinking budgets and absent recent or vividly remembered past mishaps the budget ax usually falls on risk management first. Under such circumstances it is almost inevitable that safety should suffer.

But even where organizations have a more central commitment to risk management, and even where managers take seriously their charge to meet this commitment, the nature of many risks make it difficult – perhaps impossible – to maintain optimum levels of safety. Risk events or accidents are often extremely rare.[12] For many risks demanding management attention, the probability of their occurrence is infinitesimal making it all but impossible to predict when or whether they will occur. This is especially the case for many of the risks requiring LTS.

Atrophy of Vigilance

By their very rarity, risk events are highly intermittent and often long periods of time transpire before one occurs. Circumstances such as this are characterized by the "atrophy of vigilance" (Freudenburg, 1992). While organizations and their managers may be firmly committed to exercising vigilance over risky operations, the level of vigilance, over time, atrophies. Vigilance begins at an enthusiastically high level but over time, because the offending event is so rare, that intensity (in attention and resources) seems misplaced: after all, since nothing has happened for a very long time things seem to be in order. Through complacency and boredom the level of vigilance continues to decline until a serious event, either within the organization or a similar organization, prompts a renewed commitment to safety.

Organizations, however defined, are characterized by a division of labor, a specialization and separation of tasks (often accompanied by their simplification), and a need for coordination. This organizational reality is the source of at least two other factors that tend to amplify real risks: the attenuation of information flows and the diffraction of responsibility.

Attenuation of Information Flows

The greater the simplification and separation of tasks, the greater the number of links to be coordinated in achieving organizational goals. As the number of links grows, so does this distance between those with specialized knowledge of a technology or its risks and those at the top of the organization who are responsible for broad management decisions. Accompanying an increased distance between strategic actors in an organization is an increased likelihood for miscommunication. Indeed, it increases the chances for a total breakdown in communication. The latter seems to have occurred in the case of the Space Shuttle Challenger disaster, where the concerns of technical experts never reached ultimate decision-makers (President's Commission on the Space Challenger Accident, 1986).

This is an example, explains Freudenburg (1992), of the more general pattern of the bureaucratic attenuation of information flows: information is interpreted, shaped, and modified as it moves up the bureaucratic chain of command. In instances where the information to be passed on is "bad news" the direction of attenuation is toward mollification; the inclination is to avoid the fate of the Medieval messenger by filtering out the bad and giving it a positive spin. When the information to be transmitted involves risks, the outcome is almost assured; it will be to underestimate the real risks.[13] Management practices shaped by underestimated risks are likely to be less vigorous than warranted by the misperceived underlying reality. When the risks involve the public, as they often do, the result is to increase public risk exposure.

Diffraction of Responsibility

Another way that organizations can amplify real risks is through the diffraction of responsibility. One of the defining features of organizations, as noted above, is the division of labor along lines of specialization. A concomitant of specialized tasks is a circumscribed domain of responsibility. The operator of the machine stamping out widgets is responsible for meeting production schedules and ensuring a quality widget. She has no responsibility for the activities of other domains or for the purpose to which the widget is put. Called upon to account for why widgets are being produced faster than they can be used in the assembly department, the operator is likely to respond, "that's not my department." Asked about the use of widgets in assault rifles that contribute to urban violence, the operator is likely to respond, "that's not my responsibility." The responsibilities of other members of the organization are similarly circumscribed. Thus, viewed from above, the organization can be seen as a collection of distinct multiple domains of responsibility, with management as the coordinating agent.

Structural Secrecy

The links between the separate domains may be weak or, worse, may be totally unconnected. This is particularly true of large, complex organizations charged with managing technological systems, resulting in what Vaughan (1996) has called "structural secrecy." The number of possible combinations among links increases exponentially with the addition of each new link. With a marked increase in the number of possible linkages, the ability for organizations to ensure that all important linkages are in fact "linked" becomes ever more problematic. Indeed, a complete assurance may be impossible. A proper connection among links is especially crucial when safety is at stake. Given their sheer number alone, it is easy to overlook key safety links in technological systems, thereby enhancing the possibility that something could "fall through the cracks." The consequence is again to amplify real risks.

Selecting organizations as risk actors also highlights the fact that they, as well as larger institutions, are culpable – indeed, more culpable than is commonly assumed. One sociological feature of the evolution of societies into modernity and postmodernity is the growth in what Weber (1958 [1919]) called "intellectualized rationalization" where institutions acted according to principles and calculation, rather than tradition and magic. Another sociological feature is growth in the division of labor. The latter growth is evident both within and between organizations as well as among the larger institutions of modern societies. Organizations and institutions are increasingly responsible for fulfilling wants and needs once accomplished via self-reliance and for achievements beyond the capacity of individuals working alone. But accompanying the decline in self-reliance is the growth of interdependence. The net result of these two countervailing forces is a troublesome paradox: "... the very division of labor that permits many of the achievements of advanced industrial societies may also have the potential to become one of the most serious sources of risk and vulnerability" (Freudenburg, 1993, p. 912).

Actions necessary to protect against growing risks and vulnerability may, just as with organizations, fall through institutional cracks. Whether their source be individual actors, organizations, or entire systems, technological failures are increasingly likely to happen. The source of such failures may lie with the increased potential for *recreancy*, "...behaviors of persons and/or of institutions that hold positions of trust, agency, responsibility, or fiduciary or other forms of broadly expected obligations to the collectively, but that behave in a manner that fails to fulfill the obligations or merit the trust" (Freudenburg, 1993, pp. 916–917). The larger sociological implication of this insight is that with a growing number of risks and vulnerabilities there is a growing chance for failure. And these chances increase with time, especially over the lengthy periods associated with stewardship. This, in turn, is likely to lead to a growth in recreancy that is perceived by members of society. Should this obtain, it could produce a risk of another sort and of another magnitude: risk to the social fabric itself (Short, 1984). A widespread belief in recreancy threatens the trust and mutuality upon which social relations – indeed, upon which society – are founded.

CONTEXT AND POWER

A first principle of sociology is that context matters. But so does power. Indeed the importance of context rests largely on the fact that it provides the settings for the exercise of power. Power, by its very nature, entails inequality in prerogatives and resources. Risks spur decision-making within organizational and institutional settings, but because the very items or processes at risk are often ambiguous the decision-making context is protean. Elites exercise their power by imposing concreteness on malleable contexts. Under such circumstances risk is, therefore, the footing that undergirds the power manifested in concreteness. It, too, provides an opportunity for the exercise of power.

Power is ignored by virtually all theoretical applications of RAP. For this and other reasons, a volume by Short and Clarke (1992) is uninhibited in its rejection of RAP as an explanatory framework. Clarke (1992), in fact, presents an explicit criticism of theories that focus on the individual, such as RAP as well as cognitive psychology, seeking to explode the canons of both. To understand risk is to understand not the individual decision-maker, but the conditions and circumstances under which decisions are made. The cognitive heuristics and biases that shape individual risk perceptions, amply demonstrated in cognitive psychology, are themselves shaped by organizational and institutional contexts. The front stage is where "flawed" risk perceptions take place, apparently due to our cognitive limitations, but the stage itself has been set with conditionalizing props by a variety of elite actors and decision processes. It is these actors and processes that beg of theoretical investigation.

The Columbia Disaster

We began our analyses with a narrative, with the story of the 1986 Challenger disaster. It seems fitting to conclude the analyses with a comparable story: the 2003 Columbia disaster.

On February 1, 2003 at 9:16 AM EST the Space Shuttle (also known as an orbiter) Columbia was to land at the Kennedy Space Center in Florida. At 8:59/15 s A.M. Mission Control Center, having noticed a handful of glitches in the temperature sensors and a loss of pressure in the left-side tires of the shuttle, sent the following message to the shuttle crew: "And Columbia, Houston, we see your tire-pressure messages and we did not copy your last call." At 8:59/32 s, in the typical calm voice of space discourse pioneered by Chuck Yeager, the shuttle's commander Rick Husband replied, "Roger, uh ... ". Then, nothing. This was the last communication between shuttle and command center before the Columbia, traveling at 12,700 miles/h at 200,000 ft above the earth with its surface heat at 3,000 degrees, ignited into a massive fireball, scattering debris from its disintegration over two states – rural northeastern Texas and the tip of western Louisiana.

Seven astronauts, including the first Israeli in space, perished.

This was the 112th shuttle mission since NASA had begun the shuttle program in 1981. With the launch of the Columbia in January 2003, all but one of the previous flights had been successful. Columbia, the oldest of the remaining four shuttles, had experienced 27 successful previous flights. Hence, its return from its 16-day scientific flight was presumably routine, barely engaging the media or American public.

It was later determined, by the Columbia Accident Investigation Board (CAIB), that the cause of the disaster was straightforward (CAIB, 2003). It had suffered a technical failure: a piece of insulating foam from a structure connecting the shuttle to its external fuel tank, and essential to prevent burn-up, had dislodged. It then flew at 545 mph into the shielding on the edge of shuttle's left wing where it poked a hole that allowed super hot gas into the wing upon re-entry, causing the shuttle to break apart. But, why did the foam fly off in the first place?

Lessons Learned from Space Shuttle Disasters?

What lessons for LTS on earth can be drawn from disasters two decades apart in space? The two space disasters provide a unique "natural" experiment for judging the effectiveness of management correctives introduced after the first disaster, Challenger. The two national commissions that examined the Challenger disaster not only diagnosed its cause but also made recommendations for new management practices. Key among these recommendations was the need to perform formal risk analyses.[14] After some delay, NASA did perform such analysis, using PRA, the quintessential example of RAP, completing the study in 1995. Presumably, the PRA would give signals of unacceptably high risks for future space flights, while pin-pointing the weak hardware component or human action requiring correction. That a second disaster occurred eight years after the analysis suggests that PRA was an inadequate tool for detecting and reducing mission risks.[15]

Diane Vaughan's (1996) analysis came to a conclusion about the Challenger disaster at great odds with the commissions' reports. The problem did not lie in specific human actions or amoral intentions, but in a NASA culture that trumped engineering analysis with management judgment, that allowed considerable safety slippage over time (perhaps an atrophy of vigilance), and perhaps subordinated safety to other agency goals. These lessons were lost on NASA. Rather than modifying its organizational culture, NASA chose to conduct business as usual – augmented with RAP-based risk analysis.

It took a second disaster, the Columbia, before the insights of organizational analyses, such as those provided by Vaughan, would surface again. The report analyzing the disaster by the CAIB first recognized the limitations of RAP directly, and PRA indirectly when it observed that:

> Many accident investigations make the same mistake in defining causes. They identify the widget that broke or malfunctioned, then locate the person most closely connected to the technical failure. When causal chains are limited to technical flaws and individual failures, the ensuring responses aimed at preventing a similar event in the future are equally limited ... (CAIB, 2003, p. 177).

The report also underscored – consistent with the argument developed here – the importance of understanding the organizational features of NASA, especially its culture.

> In the board's view, NASA's **organizational culture** and structure has as much to do with this accident as the external tank foam. Organizational culture refers to the values, norms, beliefs, and practices that govern how an institution functions. At the most basic level, organizational culture defines the assumptions that employees make as they carry out their work (CAIB, 2003, p. 177) (emphasis added).

Most extraordinary about this conclusion is that it is nearly a *verbatim* replication of Diane Vaughan's conclusions seven years earlier.[16] Indeed,

Vaughan, in testimony before the CAIB (CAIB, 2003), commented that:

> NASA never fixed the underlying institutional problems that led to the 1986 Challenger disaster and many errors were repeated during Columbia's doomed flight ... a similar culture at the space agency during both accidents hindered free thinking and truly open communication.

The proximate lesson to be learned is that despite the considerable resources at its disposal, despite employing the nation's best and brightest, and despite analyses showing how NASA could be the High-Reliability Organization to which it aspires, NASA failed this potential. In particular, NASA failed to meet the criteria of HROs: a commitment to safety at the top levels of management, redundancy in hardware and personnel, development of a safety culture, and organizational learning through trial and error. The implication for LTS is disturbingly straightforward: if the agency responsible for the nation's most glamorous technological accomplishments proves to be unreliable, what about some of the least glamorous technological challenges – the management of the nation's technological garbage?

The more general lesson from the space disasters for LTS was recently provided by Craig (2004), who argues that the mistakes made by NASA leading up to the Challenger and Columbia accidents are generic and characteristic of all complex institutions facing major risk management challenges. In particular, he argues that the lessons of NASA apply to the most challenging of all LTS problems, the sequestering and caretaking of high-level nuclear wastes (HLW) – whose safety must be assured for 10,000 years. DOE, the federal agency charged with disposing of HLW, has an organizational structure and culture that mirrors NASA's. It is schedule-driven, has insufficient resources to complete its job properly, is fragmented with ineffective communication, is hierarchical, and, due to hubris, is unresponsive to criticism thereby obviating its ability for trial-and-error learning.

IMPLICATIONS FOR LONG-TERM STEWARDSHIP

The principal point of this paper has been to argue for an expanded role for organizational analyses of risk and, in particular, for more work on organizational risk management. Its instrumental goal was to draw lessons from the risk literature for the management of LTS. The approach has not been toward theory construction or development per se, but toward developing an argument for a broadening of the narrow perspective, RAP, that now overwhelmingly dominates the risk field. Embedded within this point is the premise that theoretical and management questions about risk in general, and LTS in particular, cover too broad an intellectual landscape to yield to a single conceptualization. And the catastrophic consequences of many risks demand our best efforts toward understanding them. The potential for massive physical damage and financial loss, the risk to the health of first-, second-, third-, and fourth-party victims, demand our best efforts at understanding how to manage risks over the long term.

Approaching risk from the singular perspective of RAP deflects attention from organizational and institutional actors and toward individuals. One consequence of this perspective, as we have seen above, is to create a set of explanatory claims that are predestined to "blame the victim." Humans are the cognitive components in complex systems. System failures can be traced to these components – to operators, to inspectors, to managers. This line of reasoning, by making individuals culpable, is blind to the possibility that it is the system itself that is at fault and, therefore, culpable. Inadequately designed or poorly suited systems just about ensure that the humans operating the system will commit errors. It is unrealistic to expect that individuals – even the most competent and dedicated among us – can overcome systems that are poorly designed and destined to fail.

Also embedded within our main point is the premise that intellectual vitality and policy creativity are enhanced when a dialectic of perspectives is ensured – when multiple perspectives are brought to bear on common problems and issues. The field of risk can, therefore, be viewed as comprising multiple domains. Understanding each of these domains calls for a larger analytical toolbox. Here, the case was made for organizations as one of these domains, and organizational analysis, via methodological holism, as the appropriate theoretical tool. The case for the importance of organizational analysis of risks was built around a pedagogical exposition of existing work in this emergent perspective. The work reviewed, still in its incipient stages, has left many gaps. Indeed, it has left as many, perhaps more, questions than answers. Nevertheless, also revealed were a compelling set of insights demanding further scrutiny and test. The risk field will be well served with further growth in organizational analysis that engages these insights in order to refine this underutilized theoretical tool. LTS would be equally well served to capitalize on these insights.

NOTES

1. There is astounding variation in the size, type, and risks of the broad range of waste sites. That being so, it embeds key challenges: When has cleanup proceeded as

far as it could and stewardship should, therefore, begin? Will the site be amenable to or require additional cleanup as new technologies are developed? Will engineered barriers be sufficient to protect public health and the environment? And can existing institutional controls work effectively, or are new controls necessary?

2. DOD is responsible for cleaning and then transferring thousands of military facilities, such as firearm and bombing ranges; DOE is responsible for cleanup and stewardship of the nuclear weapons complex as well as high and low-level nuclear waste elsewhere; DOI is responsible for the polluted runoff from thousands of mines on Federal lands; EPA is responsible for overseeing the cleanup and stewardship of thousands of hazardous waste sites on both federal and private lands; and NASA is responsible for the decommissioning of research reactors and the cleanup and stewardship of rocket propulsion contamination.

3. The Report of the Congressional Commission was even more unrestricted in its blaming of individuals, stopping just short of naming names.

4. For a brief historical account of those roots see Jaeger, Renn, Rosa, and Webler (2001).

5. An elaboration and detailed critique of the axiomatic formulation of RAP can be found in Jaeger et al. (2001).

6. Closer examination of the Kemeny Report, however, reveals that human factors were only one factor in the accident: organizational and management practices were of key importance and specified as well, but these factors were virtually ignored by the media.

7. Coleman (1993, p. 2) traces the origination of the term to the Italian jurist who later became Pope Innocent IV, where the term, translated as a "fictitious person," defined an organization's legal standing to be equivalent to that of a natural person (the *persona naturale*) even though the former lacks a "physical corpus."

8. A modified version of this premise directs the research of a group of scholars looking at complex organizations charged with managing risky technologies that are highly reliable, termed high-reliability organizations (HROs).

9. Although Perrow's approach is clearly structural, it does draws on one type of reductionism – cognitive psychology – especially the findings on heuristics and biases and on laypersons perceptions of risks.

10. But, Perrow (2004) does not view Chernobyl as a normal accident.

11. A pioneering effort in this direction was undertaken by political scientist Scott Sagan (1993), who examined the system responsible for the safety of nuclear weapons. Because there has never been an accidental or unauthorized detonation of a nuclear weapon, the US nuclear weapons system seemed to resemble an HRO, the type of system Sagan expected to find. But once he shifted his attention to "close calls," the weight of evidence led him to the exact opposite conclusion: that the NAF better accounted for the variety of dangerous incidents he examined.

12. A fact even undisputed by the NAF.

13. Nobel laureate Richard Feynman asked, as a member of the Rogers Commission, what the risk is of each shuttle mission. NASA engineers estimated 1 in 100, while NASA estimated 1 in 100,000, a three order of magnitude difference.

14. Up to the time of the Challenger disaster NASA had literally conducted no formal risk assessments. Its technical evaluations relied upon the best engineering practices and simulations.

15. The PRA estimated the catastrophic failure of shuttle flights to be one in 145 flights. With the loss of Columbia, the actual failure rate is 2 in 112 flights.

16. Said Sally Ride, former astronaut and the only expert to serve on both the Rogers Commission and CAIB, "There were echoes of Challenger in Columbia" (CAIB, 2003, p. 195).

ACKNOWLEDGMENTS

Previous versions of this paper have benefited from the comments of Robert Brulle, Thomas Dietz, Gregory Guagnano, Darcy James, Valerie Jenness, Mike Mtika, Jim Rice, Jim Short, and Paul Stern, and from invited presentations at George Mason University; Institute of Social Ecology, Vienna; the University of Klagenfurt, Austria; King's College, London; and Science Policy Research Unit, University of Sussex.

REFERENCES

Ackerman, F., & Heinzerling, L. (2004). *Priceless: On knowing the price of everything and the value of nothing*. New York: The New Press.

Clarke, L. (1989). *Acceptable risk?: Making decisions in a toxic environment*. Berkeley, CA: University of California Press.

Clarke, L. (1992). Context dependency and risk decision making. In: J.F. Short Jr., L. Clarke (Eds), *Organizations, uncertainties, and risk*. (pp. 27–38). Boulder, CO: Westview Press.

Cohen, M. D., March, J. G., & Olsen, J. (1972). A garbage can model of organizational choice. *Administrative Science Quarterly*, *17*, 1–25.

Coleman, J. S. (1993). The rational reconstruction of society. *American Sociological Review*, *58*, 1–15.

Columbia Accident Investigation Board (CAIB). (2003). *Board report* (Volume I). Washington, DC: National Aeronautics and Space Administration and U.S. Government Printing Office.

Craig, P. P. (2004). Rush to judgment at Yucca mountain. *Science for Democratic Action*, *29*(2), 10–16.

Freudenburg, W. R. (1992). Nothing recedes like success? Risk analysis and the organizational amplification of risks. *Risk: Issues in Health and Safety*, *3*, 1–35.

Freudenburg, W. R. (1993). Risk and recreancy: Weber, the division of labor and the rationality of risk perceptions. *Social Forces*, *71*, 909–932.

Jaeger, C., Renn, O., Rosa, E. A., & Webler, T. (2001). *Risk, uncertainty, and rational action*. London: Earthscan.

Kahneman, D., & Tversky, A. (1979). Prospect theory: An analysis of decision under risk. *Econometrica*, *47*, 263–291.

LaPorte, T.R. (1998). The United States air traffic system: Increasing reliability in the midst of rapid growth. In: R. Mayntz, T. P. Hughes (Eds), *The development of large technological systems* (pp. 215–244). Boulder, CO: Westview Press.

LaPorte, T. R., & Consolini, P. M. (1991). Working in practice but not in theory: Theoretical challenges of high reliability organizations. *Journal of Public Administration Research and Theory*, *1*(1), 19–48.

March, J. G. (1978). Bounded rationality, ambiguity, and the engineering of choice. *Bell Journal of Economics, 9*, 587–608.

March, J. G., & Simon, H. A. (1958). *Organizations*. New York: Wiley.

Mintzberg, H. (1973). *The nature of managerial work*. London: Harper and Row.

Morone, J. G., & Woodhouse, E. J. (1986). *Averting catastrophe: Strategies for regulating risky technologies*. Berkeley, CA: University of California Press.

Pastina, B. (2004). Implementing long-term stewardship: A national challenge. National Academics workshop on long-term stewardship. March 16, 2004. Washington, DC. Available at http://www.cistems.fsu.edu/PDF/pastina.pdf

Perrow, C. (1984). *Normal accidents: Living with high-risk technologies*. New York: Basic Books.

Perrow, C. (1986). *Complex organizations: A critical essay* (3rd ed.). New York: Random House.

Perrow, C. (1999). *Normal accidents: Living with high-risk technologies* (2nd ed.). Princeton: Princeton University Press.

Perrow, C. (2004). A personal note on normal accidents. *Organization and Environment, 17*, 9–14.

President's Commission on the Accident at Three Mile Island. (1979). *The need for change: The legacy of Three Mile Island*. Washington, DC: U. S. Government Printing Office.

President's Commission on the Space Challenger Accident. (1986). Report of the presidential commission on the space shuttle Challenger accident (Vol. I. II, III, IV, V). Washington, DC: U.S. Government Printing Office.

Rasmussen, N. C. (1981). The application of probabilistic risk assessment techniques to energy technologies. *Annual Review of Energy, 6*, 123–138.

Roberts, K. H. (1990). Some characteristics of one type of high reliability organization. *Organizational Science, 1*, 165–170.

Roberts, K. H. (Ed.) (1993). *New challenges to understanding organizations*. New York: Macmillan.

Sagan, S. D. (1993). *The limits of safety: Organizations, accidents, and nuclear weapons*. Princeton: Princeton University Press.

Scott, W. R. (1992). *Organizations: Rational, natural, and open systems* (3rd ed.). Englewood Cliffs, NJ: Prentice-Hall.

Short, J. F., Jr. (1984). The social fabric at risk: Toward a sociological transformation of risk analysis. *American Sociological Review, 49*, 711–725.

Short, J. F., Jr., Clarke, L. (Eds). (1992). Organizations, uncertainties and risk. Boulder, CO: Westview Press.

Simon, H. (1957). *Models of man*. New York: Wiley.

Slovic, P., Fischhoff, B., & Lichtenstein, S. (1977). Behavioral decision theory. *Annual Review of Psychology, 28*, 1–39.

Starr, C. (1969). Social benefit versus technological risk: What is our society willing to pay for safety? *Science, 165*, 1232–1238.

Tenner, E. (1996). *Why things bite back: Technology and the revenge of unintended consequences*. New York: Alfred Knopf.

Tversky, A., & Kahneman, D. (1981). The framing of decisions and the psychology of choice. *Science, 211*, 453–458.

Vaughan, D. (1996). *The challenger launch decision: Risky technology, culture and deviance at NASA*. Chicago: University of Chicago Press.

Weber, M. (1998 [1919]). Science as vocation. In: H. Gerth, & C. W. Mills (Eds), *From Max Weber: essays in sociology*. New York: Oxford University Press.

Wildavsky, A. (1988). *Searching for safety*. New Brunswick, NJ: Transaction Books.

Printed and bound by CPI Group (UK) Ltd, Croydon, CR0 4YY
16/09/2026
14963402-0005